Interkosmos

The Eastern Bloc's Early Space Program

Colin Burgess • Bert Vis

Interkosmos

The Eastern Bloc's Early Space Program

 Springer

Published in association with
Praxis Publishing
Chichester, UK

Colin Burgess
Bangor, New South Wales, Australia

Bert Vis
Den Haag, The Netherlands

SPRINGER-PRAXIS BOOKS IN SPACE EXPLORATION

Springer Praxis Books
ISBN 978-3-319-24161-6   ISBN 978-3-319-24163-0   (eBook)
DOI 10.1007/978-3-319-24163-0

Library of Congress Control Number: 2015953234

Springer Cham Heidelberg New York Dordrecht London

Front cover: The Soyuz-30 Interkosmos crew of Pyotr Klimuk and Miroslaw Hermaszewski.
Rear cover: (1) English version of the Interkosmos patch; (2) Dzhanibekov and Gurragchaa of Soyuz-39; (3) Gubarev and Remek (Soyuz-28) at the launch pad; and (4) Dumitru Prunariu aboard Salyut-6
Project copyeditor: David M. Harland
Cover design: Jim Wilkie

Printed on acid-free paper

Springer International Publishing AG Switzerland is part of Springer Science+Business Media (www.springer.com)

Contents

Dedication ix
Foreword x
Acknowledgements xii

1 History and development of the Interkosmos program 1
    A bold international venture 1
    The early days of Interkosmos 2
    The first Interkosmos satellite 4
    Research program continues 7
    Expanding Interkosmos 9

2 First to fly: Czechoslovakia's cosmonaut 11
    A cooperative venture 11
    The right place at the right time 13
    Surviving the selection process 14
    Training begins 16
    First Czech in space 21
    Living and working in orbit 24
    Home to adulation 27
    A time of danger 29
    Later accomplishments 31

3 From Poland to Salyut-6 35
    Tiny survivor 36
    Space flight competitors 38
    Second candidate 43
    Chosen to fly 44
    Rendezvous with a space station 47
    Safely back on Earth 51

4   Sigmund Jähn and Soyuz-31 57
A golden opportunity 60
Training in Star City 62
Ahead of lift-off 65
Rendezvous, docking and hard work 66
Back home a hero 72

5   Bulgaria in space 77
A dream of flight 77
A whole new direction 79
The making of a cosmonaut 83
Docking failure 84
Reluctant homecoming 86
One and only flight 88

6   Hungary joins the space club 92
A shoemaker's son 93
Over to Moscow 94
And then there were two 95
Cosmonaut training 98
On to Baykonur 101
Heading for Salyut-6 103
Return to Hungary 106

7   A Vietnamese cosmonaut 110
Born into poverty 110
A second Red Star pilot 113
The final two 114
Concentrated training 116
An Olympic propaganda mission 119
Working in orbit 120
Homeward bound 122

8   The Cuban Salyut mission 127
A child of Cuba 127
Becoming a pilot 128
The call of the cosmos 129
Joint Soviet-Cuban mission 136
Life after space 139

9   From the steppes of Mongolia 143
Cosmonaut candidates 143
Testing times 145

Flight into the cosmos 149
Working aboard the space station 150
The journey home 153
Later life 154

10   Romania continues the program 159
"I saw another future" 159
Initial rejection 161
Star City 163
The final choice 166
A Romanian in space 170
Fall from grace 174
Back in the air 177
Positions of eminence 179

11   Beyond Interkosmos: Soyuz T-6 184
Child of the war 184
Selecting a spationaute 186
Out of Africa 188
Training regime 189
After the first year 190
Occupying Salyut-7 194
Back home 198
Frenchman on a space shuttle 200

12   Soviet-Indian mission 205
A joint flight is proposed 205
The sound of jets in the sky 206
Fighter pilot ambitions 208
Cosmonaut training center 210
Training intensifies 212
Activity aboard Salyut-7 215
Research continues 218
End of a mission 219
Sharma looks back 221

13   A Syrian researcher on Mir 224
Two men of Syria 225
Crew assignments and training 225
Official announcement 227
Plans for the mission 228
Syria in space 230
A hitch on the way home 233

14   Bulgaria's second flight 238
    A new cooperative science mission 238
    Aleksandrov back in training 239
    Bulgaria's second flag in orbit 242
    Mission to Mir... 244
    A multitude of awards.. 246
    Results of a successful flight.. 250

15   Afghanistan's cosmonaut-researcher 252
    Training for the skies 252
    Reporting for duty.. 254
    Link-up with Mir.. 258
    Emergency landing 260
    Post-flight celebrations.. 262

16   Chrétien in space again 266
    A second trip into space 266
    Training for the Aragatz mission 270
    A shortened stay... 271
    Returning to space.. 274
    A French spacewalk 276
    Working aboard Mir.. 278
    Landing and recovery 279
    Life after space... 281

17   The Interkosmos missions in retrospect.................................... 285

18   Philately and the Interkosmos program by James Reichman 291

Appendices... 308

About the authors 312

Index... 314

Of necessity, two international guest pilot/ researchers were chosen for each of the Interkosmos manned missions. This meant that while one would know the glory of flying into orbit and representing their nation, the backup pilot would be forever consigned to living in their colleague's shadow. Although they contributed much to the success of the Interkosmos program, their brush with fame is far less known. This book, therefore, is respectfully dedicated to those who remained behind; whose dreams and ambitions of flying and working amid the magnificence of space fell sadly short.

Oldřich Pelčák (Czechoslovakia)
Zenon Jankowski (Poland)
Eberhard Köllner (German Democratic Republic)
Béla Magyari (Hungary)
Bùi Thanh Liêm (Vietnam)
José Armando López Falcón (Cuba)
Maidarzhavyn Ganzorig (Mongolia)
Dumitru Dediu (Romania)
Ravish Malhotra (India)
Munir Habib (Syria)
Krasimir Stoyanov (Bulgaria)
Mohammad Dauran (Afghanistan)

Foreword

As a child, I could never have imaged that one day I would fly into the vastness and glory of space. Now, as I look back, I am grateful to my parents and teachers who supported and educated me in my early years and who impressed upon me that anyone can achieve their dreams through trial and perseverance.

I recall in my youth being fascinated by the amazing stories of French author Jules Verne. Through his words he carried me in delicious excitement to the depths of the ocean and on spacecraft bound for other worlds, which I longed to see for myself. He transported me to the Moon and back in the most unbelievable manner and served as a wonderful inspiration.

I was just eight years old when Yuriy Gagarin became the first human being to fly into space in April 1961. As I listened to the sensational news reports of his flight on Radio Romania, I realized that this was a truly momentous event; one that would forever change the world in which we lived. We were on our way to the stars. I felt, like many others, that his flight had offered me a whole new dream. Little did I know back then that just over twenty years later I would follow Gagarin into orbit as the first Romanian to enter that magnificent, starry domain, becoming the 103rd explorer of outer space.

It is a pity for the sake of humanity that not everyone can fly into orbit and look down on our glorious blue planet. It certainly has the power to change one's attitude, not only by the extraordinary beauty of the Earth, but by how truly fragile it is, and how we must ensure that we preserve it for future generations. Looking at the Earth from space is like witnessing a fascinating symphony of colors, from white and gray hues through to brilliant shades of green, blue and brown. Those images will always stay within me.

It is easy to become mesmerized by the magnificence of our planet seen from orbit, but we are also venturing ever deeper and staying ever longer in space in our ongoing quest for knowledge. Human knowledge certainly has no limits. Today, aboard the orbiting station, men and women are pursuing international, multilateral scientific activities in such fields as medicine, biology, meteorology, and the cosmic environment. In the early days of space exploration epochal discoveries were made on every flight, most of which we now take for granted. These days, discoveries are of greater refinement, but we need to continue our quest for information about who we are, how best to look after our planet and its people, and where we should venture in the future. Among other things, future generations will hopefully come to a better understanding that the balance and harmony in the universe should be maintained around us. Beauty and nature reserves, which we still have here on Earth, must be kept for those who come after us.

There is an old adage about being in the right place at the right time, and that surely happened for me. I was the right age, living in the right era, and with the appropriate skills and qualifications to become a candidate to fly into space for my country. I had kept my childhood dream alive, and I will always encourage young people to do the same thing.

Dumitru Prunariu in 2013.

It is personally exciting that the two authors, both known to me, have embarked on writing this book devoted entirely to the Interkosmos program. The reader should be pragmatic and realize that although much of what was created and carried out in that program was beneficial and in the name of international cooperation and scientific endeavor, like many programs of the space powers of those times it also had its propagandist elements. But we were all swept up in the excitement of what we were doing as cosmonaut-researchers, in the knowledge that we were being offered an unbelievable opportunity which was available to only a handful of people.

The authors have no agenda in writing this book other than to present and preserve the remarkable history of the Interkosmos program, and the stories of the men, like me, who flew into space as proud representatives of their nations. I wish the authors well and support them in this endeavor, and I am truly delighted to have been able to contribute to making this book a reality.

Dumitru-Dorin Prunariu,
Romanian Cosmonaut-Researcher, Soyuz-40

Acknowledgements

Over the course of many years of friendship, and with their shared passion for recording the history of human space flight, the authors have often discussed collaborating on a book that would apply their accumulated knowledge of Soviet/Russian space history and document an important aspect of that momentous story. The Interkosmos story is one that has never been fully told in book form, although it has featured in many publications and a handful of non-English autobiographies penned by the participants from former Soviet bloc nations. So this book is that long-overdue collaborative effort.

The authors have mined a great many publications in several languages for first-hand accounts of the manned Interkosmos flights. However a major feature of this book is the number of personal interviews with the Interkosmos participants conducted over many years by Bert Vis, sometimes solo, but also in the company of several dedicated space sleuths and historians, principally the late and sadly-missed Rex Hall, MBE, widely regarded as the doyen of Soviet space historians. A great friend to both of the current authors, Rex's decades of dedication to the subject and trips to Moscow's Star City yielded a veritable mountain of information in the form of documents, magazines, newspapers and photographs, to which we were kindly given access by his wonderful wife, Lynn Hall. The sources of many of the rare photographs used in this book could not be identified, so we have been unable to provide full accreditation. Those that were taken before the fall of the Soviet Union are not encumbered by copyright. We regret that some photographs do not bear any accreditation beyond that of having come from either our own files or from those of Rex Hall, but we will be delighted to accept messages offering correct credit notations and will endeavor to have these inserted in subsequent editions of this book.

In putting these acknowledgements together, the mere listing of names can hardly express our profound and ongoing thanks to those who so willingly assisted in the compilation of this book. We are very grateful and thankful for their kind assistance and support in bringing this publication to reality.

Of the Interkosmonauts who responded to our recent requests for additional information, special thanks go to Dumitru-Dorin Prunariu (Romania) and Zhugerdemidiyn Gurragchaa (Mongolia), and former GDR candidate Eberhard Golbs. We also gratefully acknowledge the kind assistance of Bart Hendrickx, Jürgen Esders and Lida Shkorkina, and Francis French of the San Diego Air and Space Museum.

Many of the photographs used in this book came to the authors through the courtesy of Joachim Becker, who compiles and administers the amazing Spacefacts website (available at *www.spacefacts.de*). A must for any space flight enthusiast, this website not only provides well-researched facts about every space mission and anyone ever selected to fly into space, it also offers many photographs that are not available anywhere else. We are very grateful and indebted to Joachim for giving us high-resolution photographs (where available) concerning the Interkosmos program, and also for his wonderful assistance on previous publications. This book is all the better for his support.

Ongoing thanks also go to everyone at Springer-Praxis, who have always been so easy to work with; in particular Senior Editor Maury Solomon in New York and her trusty off-sider, Assistant Editor Nora Rawn. We must also thank most profusely the work of Clive Horwood at Praxis Publishing in the United Kingdom. Bravo, Clive! Many heartfelt kudos must also go to our wonderfully diligent copyeditor and fellow space author David M. Harland, who always manages to identify and repair problem areas in the manuscript, tidy up instances of poorly constructed sentences, and spot factual errors. He gives that final polish to our work and truly deserves a stirring round of applause. And of course our appreciation goes to Jim Wilkie, who puts together the brilliant cover art of these books.

1

History and development of the Interkosmos program

In a rousing speech that he gave at a Moscow rally on 22 October 1969 to welcome back the crews of the recent 'troika flight' of Soyuz-6, Soyuz-7 and Soyuz-8, Soviet General Secretary Leonid Brezhnev observed: "Our country has an extensive space program designed for many years ahead. We are following our own road, following it consistently and purposefully. Our road of space conquest is a road of solving fundamental tasks, basic problems of science and technology. The Soviet Union regards space explorations as the great task of learning and practically using the forces and laws of nature in the interests of men of labor, in the interests of peace on the Earth."[1] These goals would be integral to a new and attention-grabbing facet of the propaganda-driven multi-national space effort that came to be known as Interkosmos.

A BOLD INTERNATIONAL VENTURE

Interkosmos was the name given to a limited international program of peaceful cooperation in space research and technology originally dating from the mid-1960s, which was devised and orchestrated by the Soviet Union. Those behind the venture described it as a means of establishing mutually beneficial relations with Eastern Bloc countries through unmanned and manned space ventures. But it also served as a high-profile propaganda exercise, with the Soviet Union playing a particularly dominant role in the manned research program which carried "guest" cosmonaut-researchers into space, and controlling the release of most of the publicity associated with that program.

There was actually a practical engineering initiative behind the origins of the manned Interkosmos program. Spacecraft designers and engineers had long agreed that a Soyuz spacecraft should spend no more than 100 days in orbit. Beyond that time, the vehicle's batteries and propellants would have gradually degraded, and this could have a potentially serious impact when crews attempted to bring the Soyuz back to Earth. This had obvious ramifications for the future Soviet space program, which involved sending cosmonauts on extended stays aboard orbiting Salyut space stations. As James Oberg pointed out in his book, *Red Star in Orbit*, they soon came up with a solution, but one that was extremely wasteful and fraught with unsatisfactory technical difficulties.

© Springer International Publishing Switzerland 2016
C. Burgess, B. Vis, *Interkosmos*, Springer Praxis Books, DOI 10.1007/978-3-319-24163-0_1

"The solution was to send up a fresh Soyuz periodically to replace the aging one. On board the ship would be mail, fresh food and other cargoes; the men aboard Salyut could use their original Soyuz to send back down to Earth the results of their experiments, including exposed film, logbooks, tape cassettes, biological and medical samples, and materials produced in the furnaces.

"Sometime in 1976 or so, planners must have realized that these Soyuz replacement flights need not be unmanned, flown entirely on autopilot … Instead, two cosmonauts could visit the space crew, cheering them up."[2]

Discussions were held as to the possible crewing on these Soyuz change-over missions, but sending two Soviet cosmonauts was seen as an unnecessary waste of resources. Instead, it was felt that some useful science could be carried out in the few days that this fresh crew would be present on the Salyut station. Thoughts then turned to sending up non-cosmonaut researchers. Obviously the commander would be a Soviet cosmonaut, and the person in the second seat need only be given minimal training to cope with any in-flight emergencies or incapacitation of the flight commander. Basically they would have to know how to control the Soyuz craft in any sort of emergency, and a better than basic knowledge of the Russian language was crucial. Discussion then began to center on what was occurring in the West. Around that time, the United States had come to an agreement with the European Space Agency (ESA), a consortium of nations including West Germany, Italy, France and Great Britain, which would allow ESA astronauts to fly aboard future space shuttle missions and participate in work programs, principally in the Spacelab module that ESA was to develop.

Although the first shuttle flight was still some years away, the Russians seized upon this initiative, which carried great potential propaganda value, and the Interkosmos program was born. It was formally announced in mid-1976, with the Eastern Bloc representatives being referred to as research-cosmonauts.

THE EARLY DAYS OF INTERKOSMOS

The Interkosmos program actually had its origins as far back as 15–20 November 1965, when designated representatives from the Soviet Union met with delegations from eight fraternal Communist countries to discuss the content, form, and objectives of a possible collaborative effort in space science and exploration. Those eight countries were the German Democratic Republic (GDR), the People's Republic of Bulgaria, the Republic of Cuba, the Hungarian People's Republic, the Mongolian People's Republic, the Polish People's Republic, the Romanian Socialist Republic and the Czechoslovak Socialist Republic. At the conference, the signatory participants exchanged views concerning the most suitable ways and means of pursuing cooperation in the exploration and peaceful use of outer space, taking into account the scientific and technical potentialities and economic resources of each particular socialist country. Questions on forming a program of joint research in the fields of space physics, space meteorology, organization of distant communications and telecasts, and space medicine and biology with the help of artificial satellites and geophysical and meteorological rockets were discussed. As well, the question of jointly constructing and launching satellites and the possibilities of specialists from interested countries cooperating in the development of new instruments and equipment for space exploration were discussed.[3]

In Moscow six months later, on 30 May 1966, a Council for International Cooperation in the Peaceful Exploration of Outer Space was established under the auspices of the Academy of Sciences of the USSR – the highest scientific institution in the Soviet Union – which was at that time led by Academician Boris N. Petrov. The Council determined that the Soviet Union and the eight participating nations should each create a body to formulate a scientific program for this undertaking.

Academician Boris Nikolayevich Petrov, President of the Council for International Cooperation, USSR Academy of Sciences. (Photo: Author's collection)

At another conference the following year, held from 5–13 April, all nine nations signed an agreement to work together on a number of comprehensive space cooperation projects in five areas of study: physics, meteorology, biology and medicine, space communications, and the study of the Earth's natural resources. The protection of the planet's environment would be added to the list in 1976. Further agreements reached at the conference covered the holding of further conferences, symposia and meetings, as well as the exchange of visiting scientists.

Sixteen months later, on 13 August 1968, the signatories to the agreement submitted to the United Nations' Secretary General a draft plan for an *INTERSPUTNIK* commercial system of international space communication that would meet the requirements of both developed and developing countries.

THE FIRST INTERKOSMOS SATELLITE

On 20 December 1968 the Interkosmos program took flight with the launch into orbit of the DS-U2-GK satellite (officially designated Kosmos-261) from the Plesetsk spaceport, located some 500 miles north of Moscow. The satellite was to study variations in the parameters of the upper atmosphere and the nature of the phenomena of auroras. Scientists from Bulgaria, Hungary, the GDR, Poland, Romania and Czechoslovakia worked with the Soviet Union to carry out simultaneous terrestrial observations with the satellite and then participate in an analysis of the results.

The Soviet political newspaper *Pravda* invited Academician Petrov, the President of the Council for International Cooperation under the USSR Academy of Sciences, to express his opinion about the mission of this particular satellite. It quoted him as saying:

"Joint work by the scientists of the socialist countries in the field of space physics has been conducted since 1957, when the first artificial Earth satellite was launched. At first this collaboration was limited to the joint optical observations of the satellite on the ground, and investigations based on those results. A new stage on this path was the joint fulfillment of scientific experiments with the help of Soviet satellites and rockets, in accordance with the program of collaboration between socialist countries in outer space which was adopted in Moscow in April 1967. One of these experiments is being undertaken using the satellite Kosmos-261 in conjunction with geophysical observations on the ground.

"The satellite has been injected into a near-polar orbit. Its launching has been timed for a period close to the maximum in the solar activity cycle. It carries equipment to measure the characteristics of geoactive corpuscles, that is, electrons and protons, and to measure the variations of the density of the upper atmosphere.

"This is a multiple experiment. It also includes coordinated ground observations at the geophysical stations of the Soviet Union and other socialist countries, and scientists from Bulgaria, Hungary, the German Democratic Republic, Poland, Romania and Czechoslovakia are taking part in these observations.

"The program envisages various investigations. It includes during the period of work of the satellite, studying the characteristics of the ionosphere by vertical ionospheric probing, measuring the absorption of radio waves in the ionosphere and studying sudden iono-spheric perturbations. The investigations in the polar latitudes include also measuring variations of terrestrial magnetic fields and terrestrial currents, as well as making photo-graphic, spectral, electrophotometric and radar observations of auroral phenomena. Moreover, changes in the orbit of the satellite will be measured, in particular during magnetic storms and auroras, as a means of calculating the density of the atmosphere.

"The multiple nature of the experiment and the use of different methods of investigation is connected with the fact that it is impossible to study sufficiently well many characteristics of the upper atmosphere and magnetosphere of the Earth – including those which are connected with magnetic storms and auroral phenomena, which cover huge areas of near-Earth space, if the investigations are conducted from any point on the Earth. For solving these problems, it is necessary to have the international cooperation of scientists who can study these planetary processes at different points on the Earth."

After providing a comprehensive technical exposition of the satellite's capabilities and characteristics, Petrov offers his concluding comments, stating that the experiments being conducted with the assistance of the Kosmos-261 satellite are "an important step toward the practical accomplishment of the program of collaboration between the socialist countries in the exploration of the physical properties of outer space. In the near future it is planned to carry out joint experiments conducted with the launching of satellites and rockets with equipment developed in a number of socialist countries."[4]

Engineers and technicians mate the Interkosmos-1 satellite to its Kosmos-2 carrier rocket. (Photo: Author's collection)

The following year, on 14 October 1969, the first solar satellite in the Interkosmos series was launched into orbit from the Kapustin Yar Cosmodrome, first established in 1946 in the Astrakhan region, between Stalingrad (now Volgograd) and Astrakhan. The Interkosmos-1 satellite, cooperatively developed between the GDR, Czechoslovakia and the Soviet Union, was designed to study short-wave radiation emanating from the Sun and its effects upon the Earth's upper atmosphere. Its data revealed the polarization of the solar X-ray radiation and helped in the study of the distribution of oxygen in the Earth's upper atmosphere.

On Christmas Day that year, a second space vehicle was launched from Kapustin Yar, with Interkosmos-2 opening the series of ionospheric satellites. The goal of this particular satellite was to investigate the parameters of the Earth's upper atmosphere and ionosphere. Its scientific experiments had been prepared by scientists from Bulgaria, the GDR, Cuba, Poland, Romania, Czechoslovakia and the Soviet Union. They yielded data on the global distribution of electron temperature and ion concentration.

On 13 October 1969 Boris Petrov addressed a meeting of international member delegates ahead of the forthcoming launch of the Interkosmos-1 satellite. (Photo: Author's collection)

At left, the Kosmos-2 rocket stands ready to launch Interkosmos-1. On the right, a museum model of the satellite. (Photos: Author's collection)

RESEARCH PROGRAM CONTINUES

Many other satellite launches would follow, with the Interkosmos program allowing Eastern bloc countries the opportunity to carry out quality but inexpensive scientific space research. The investment of each of the nations varied widely, although it was the German Democratic Republic that provided the highest level of funding. During the period from 1960 through to the mid-1970s, the GDR participated in more than fifty ventures under the Interkosmos flag.

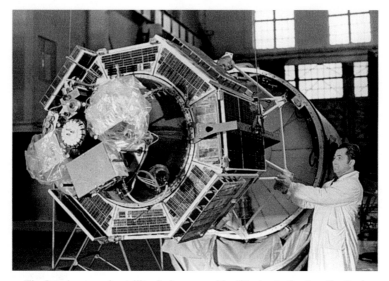

The Interkosmos-4 satellite during assembly. (Photo: Author's collection)

One series of satellites launched under the Interkosmos program was named Vertikal. In all, eleven such satellites were launched between November 1970 and October 1983, mostly carrying experiments from the Soviet Union, Bulgaria, Czechoslovakia, the GDR, Poland, Hungary and Romania to study solar short-wave radiation.

Going back in time a little, in 1972 the Soviet Union's *Opytnoye Konstruktorskoye Buro* (Experimental Design Bureau) OKB-586 had begun to develop a considerably larger vehicle known as the automatic multi-purpose orbiting station, or AUOS, which would permit the carriage and use of larger and far more sophisticated instruments and experiments. The first of these bigger satellites, designated Interkosmos-15, was launched on 19 June 1976. Two years later, Interkosmos-18 carried with it a small Czechoslovakian satellite called Magion, which separated from the parent satellite once in orbit. Two further such satellites would be launched in 1989 and 1991.

In 1981 the Interkosmos program successfully launched the Bolgaria-1300 satellite into orbit for Bulgaria. That same year France joined the program and participated in the launch of the Oreol-3 satellite, which carried a total of 12 experiments: 7 provided by France, 4 by the Soviet Union, and one as a joint venture.

The Vertikal-1 satellite atop a V-5V rocket is prepared for launch. (Photos: Author's collection)

The Magion receiving station at Panska Ves in the Czech Republic. (Photo: IAP ASCR)

A model of the Oreol-3 satellite. (Photo: Author's collection)

The break-up of the Soviet Union in 1991 led to the demise of the Interkosmos satellite program, and it was officially terminated with the launch of Koronass-1 from Plesetsk on 2 March 1994.

A full list of the Interkosmos series of unmanned satellite launches can be found in the Appendix section of this book.

EXPANDING INTERKOSMOS

Meanwhile, adding a note of confusion to the Interkosmos program, the proposed manned aspect of the international venture was also given the formal title of Interkosmos in 1970 during a summit in Wroclaw, Poland. Under this agreement, each of the eight Soviet Bloc nations would be responsible for making and supplying its own equipment, while the Soviet Union had responsibility for integrating experiments into satellites and the launching of both the satellites and manned spacecraft involved in the Interkosmos program. Adding further confusion, the Soviet committee formed to oversee these activities also went by the name of Interkosmos. This was responsible for carrying out satellite research programs for bilateral cooperation unrelated to the manned Interkosmos program. However several years would pass before the manned phase of the Interkosmos program was fully developed.

At a Moscow meeting on 13 July 1976, the governments of the nine countries involved in the Interkosmos program signed an "intergovernmental agreement of collaboration" for the peaceful exploration of space. This document authorized each of the eight bloc countries to send one of its nationals on a Soyuz mission. Moscow only required that these cosmonauts be military pilots, because their prior flight training and experience would make them better prepared for space flight.

Soviet officials then used medical and psychological criteria to set out the order in which the different nations would fly their missions. Czechoslovakia and the Poland would be the first two, then the German Democratic Republic, also known as East Germany. Given that it was the latter that participated most actively in the program, the GDR government insisted that the order be changed, but the Russians held firm.

At the same time in the United States, NASA was making preparations for the first missions of the space shuttle, announcing plans to offer places on some flights to foreign astronauts, including Europeans. The official announcement of the selection of these payload specialists, as they were known, was made on 8 July 1976, some five days prior to the signing of the Interkosmos agreement.

On the Soviet side, international flights were scheduled to begin in 1978, with a guest cosmonaut-researcher being launched in the right seat of a two-seat Soyuz to the Salyut-6 space station. The spacecraft would be under the command of a Soviet cosmonaut and the mission would last about a week. The training of the international cosmonaut would take about 18 months, and two guest cosmonauts would be trained in each case. From these a prime and alternate candidate would be selected for each flight.

In December 1976, the first six foreign cosmonauts arrived at the Gagarin Cosmonaut Training Center (*Tsentr Podgotovki Kosmonavtov*, or TsPK) in Moscow's Star City. They were the representatives of Czechoslovakia, Poland, and the German Democratic Republic.

This is the story of those manned missions into space.

REFERENCES

1. Unaccredited article, "Targets for Soviet space research," *New Science and Science Journal*, issue 6 May 1971, pp. 308–310
2. James Oberg, *Red Star in Orbit: The Inside Story of Soviet Failures and Triumphs in Space*, Random House, New York, 1981
3. G. I. Petrov (Ed.), *Conquest of Outer Space in the USSR* (translated from the Russian), Nauka Publishers, Moscow, 1971, citing *Pravda* newspaper, issue 24 November 1965
4. G. I. Petrov (Ed.), *Conquest of Outer Space in the USSR* (translated from the Russian), Nauka Publishers, Moscow, 1971, citing *Pravda* newspaper, issue 22 December 1968

2

First to fly: Czechoslovakia's cosmonaut

According to Vladimir Remek, speaking in Prague on the topic of Czechoslovakian-Soviet cooperation in the Interkosmos program on 28 May 1982, eleven years before the peaceful dissolution of Czechoslovakia into the Czech Republic and Slovakia on 1 January 1993: "Economic scientific and technological cooperation between Czechoslovakia and the Soviet Union is a steady source of values whose extent, especially in the progressive scientific fields and key technology, we can hardly estimate. Space research has a firm basis in the cooperation of the two countries today, especially as a component of international cooperation of socialist countries in the exploration and use of the universe in the Interkosmos program."

A COOPERATIVE VENTURE

In a bold initiative in late 1965 the Soviet Union announced that it would make available to all socialist countries the use of its latest space technology, as long as their participation and goals fell within the framework of peaceful exploration and use of space. Because of the complexities involved, it took time to broaden the plan and build up sufficient domestic and international momentum. As a result, the functional beginning of the Interkosmos program only took place in Moscow in April 1967. That month delegations from socialist countries met to address the basic issues of how they could cooperate in space exploration; particularly in space physics, space biology, space meteorology, and space communications. Cooperation between Soviet and Czechoslovak specialists had been evolving in this direction even before that date, both in laboratories and through Czechoslovak participation in experiments of the Soviet national space program.

Czechoslovakian participation in space exploration acquired a whole new dimension on 14 October 1969, when the first Interkosmos satellite was launched to study the Sun's radiation. Shortly afterwards, in December, the Interkosmos-2 satellite was launched to study the upper atmosphere of the Earth. In July 1976, during the signing of the new government agreement on the Interkosmos program, the Soviet delegation suggested that experiments be undertaken not only on satellites but also on manned Soviet spacecraft.

Remek continued, "Czechoslovak science participated significantly by installing and preparing experiments in all fifteen Interkosmos satellites launched thus far, as well as in 'Vertikal' geophysical rockets, the first of which was launched in the Soviet Union on

© Springer International Publishing Switzerland 2016
C. Burgess, B. Vis, *Interkosmos*, Springer Praxis Books, DOI 10.1007/978-3-319-24163-0_2

28 November 1970. Not only were there achievements in the basic research of space physics and astrophysics, biology, medicine and the like, [but] these were augmented in 1975 by geophysics and other fields when the program was expanded in a fifth direction with the remote exploration of the Earth. And the results of research in space communications led to the signing of an agreement to create the international 'Intersputnik' system and organization, which had its tenth anniversary last year.

"Manned space flights in the Interkosmos program, in which cosmonauts from socialist countries work side by side with the Soviet cosmonauts as international crews orbiting the Earth, have sometimes been called a turning point in the whole development of this program. I believe that these flights, and thus my space flight as well, were merely a logical expansion of the preceding cooperation in the peaceful exploration of the space. This soon reached the stage that even socialist countries which until then had been using only satellite technology, were able to prepare experiments for manned flights because the Soviet Union generously offered us the space in the spacecrafts and aboard the space stations."[1]

Czech cosmonaut/researcher Vladimir Remek. (Photo: Hospodarske Noviny)

THE RIGHT PLACE AT THE RIGHT TIME

Nearly thirty years after he entered the world on 26 September 1948, Vladimir Remek became the first person to travel into space who was not from the United States or the Soviet Union, with the additional honor of being the first Czech citizen to orbit the Earth.

The son of Slovakian-born fighter pilot Jozef Remek (later a lieutenant general in the Czechoslovak Army and Deputy Defense Minister of Czechoslovakia) and his wife Blanka, Vladimir Remek was born in southern Bohemia. His birthplace, which he loved exploring with his sisters Jitka and Dana, was the picturesque city of Ceske Budejovice, located near the Austrian border at the confluence of the Vltava (Moldau) and Malse rivers.

Ceske Budejovive, the capital of South Bohemia, has preserved its historical character over the centuries and today boasts one of the largest squares in Europe. (Photo: Ceske Budejovice Tourist Information Center)

Ceske Budejovice, better known abroad by its old Germanic name of Budweis, today serves as an important hub in the Czech brewing industry. The town's historical core, named for its founder King Premysla Otakara II, is one of the largest and certainly most photogenic market squares in Bohemia, with camera-bearing tourists being drawn to the visual spectacle of its baroque Town Hall, St. Nicholas Cathedral, Samson Fountain, the ornate Vcela Palace and 236-foot Black Tower, with its high lookout and belfry, dating from the 16th century.

At the age of six, Remek was taken on his first airplane flight in a Fieseler Fi-186 Storch, traveling from his hometown to Bratislava, Slovakia. While growing up he enjoyed reading aviation stories and building model airplanes, but had his youthful heart set on becoming a salesman in a pet shop. He said in one interview that he had a fascination with ornamental fish and thought it was a good opportunity to bring a few home as pets. Due to his father's career in the Czech Air Force, the Remek family moved frequently. After taking his first years of elementary education in his hometown, from Grade 7 on Remek attended school in Brno, Moravia.

There was great excitement in Brno on 12 April 1961, when the students heard over the school radio that a Soviet cosmonaut named Yuriy Gagarin had successfully completed an orbital flight around the world.

At this time Remek was also active in the Pioneers, an Eastern Bloc version of the Boy Scouts, and later participated in the Czechoslovak Youth Organization. He took his high school education in Caslavi, eastern Bohemia, majoring in physics and mathematics. As a teenager, he won many local and regional awards in 400-meter, 800-meter, and 1,500-meter events as a member of his high school track team. After graduating in 1966 he rejected his earlier aspirations of either becoming a pet shop salesman or pursuing a career in nuclear physics and decided instead to enlist in the Czechoslovak People's Army in order to learn to fly. He subsequently entered Vyssi Letecke Uciliste (a military flying college) in Kosice, near the Ukrainian border, where he first soloed in an Aero L-29 Dolphin jet trainer.

In 1969, Remek moved on to Prerov Air Base near Brno for advanced training in the supersonic Mikoyan Gurevich MiG-21 fighter. The following year he returned to Ceske Budejovice as a second lieutenant to fly the MiG-21 in the 1st Fighter Interceptor Regiment. At the age of 24, now the best pilot in his squadron, he was awarded a military scholarship and proceeded to the Gagarin Air Force academy in Russia. Here, from 1 September 1972, he took on advanced studies over the next four years. He returned home in June 1976 and five months later achieved the rank of captain. Remek then served as Deputy Squadron Commander of the 1st Fighter Squadron until he was selected for cosmonaut training. He recalls how he came to be selected to carry the Czechoslovakian flag into space.

"It was actually a rather short but very interesting and dynamic journey. Czechoslovakia in that era was part of the Eastern Bloc and was a member of an organization called Interkosmos. Other members were for example [the] German Democratic Republic (East Germany), USSR, Poland and others …. It was 1976, about a year after the first joint USSR and American project – the Soyuz Apollo, when two spacecraft docked together in orbit and the Soviet and American astronauts met in space for the first time in history. And after that, the USSR offered to the Interkosmos countries an opportunity to participate on Soviet human space flight missions.

"For me it was really something. I was just about to finish the Gagarin Air Force Academy near Moscow and so I joined this race to get that opportunity. It was very much a matter of being in the right place at the right time and having studied at the very right school. There was of course a long process of selection and at the end there were representatives from Czechoslovakia, Poland and East Germany. And in that order we all went to space in 1978. That I, as a Czechoslovak, was the first was an interplay of many circumstances."[2]

SURVIVING THE SELECTION PROCESS

The selection process began in the summer of 1976, with the Kremlin anxious to show the world a friendlier face by launching a new propaganda campaign. It was determined that the Gagarin Cosmonaut Training Center (*Tsentr Podgotovki Kosmonavtov*, or TsPK) located in Moscow's Star City did not have the capacity at that time to train any more than six foreign candidates for space flight. As each of the initially nominated Eastern Bloc countries had to supply two trainees, the first set of candidates came from Czechoslovakia, Poland and the German Democratic Republic. Other nations would follow.

The first phase in the selection of a pair of suitable Czech candidates was undertaken by the Prague Institute of Aviation Medicine. They went through the personal files of all 179 serving Czech Air Force military pilots aged 25–35 years who met the general require-ments for flying supersonic aircraft, and reviewed the results of their annual medical exam-inations. The top 100 were selected for national medical examinations. During these, the Chairman of the Standing Medical Committee, Col. (Dr.) Antonin Dvorak said he initially excluded all the physically unfit pilots; those for example, who were excessively over-weight, or suffered from conjunctivitis, ulcers, neurosis, liver inflammation, or urinary tract infection. At the end of this process, 24 names remained as possible candidates. After a far more thorough physical and psychological examination conducted at the Institute that involved further testing of the vestibular system that maintains a sense of balance, the list had been winnowed down to just eight: Frantisek Pavlik, Ladislav Klima, Stefan Gombik, Vladimir Remek, Pavol Bialon, Jan Hasan, Oldrich Pelcak and Michael Vondrousek. The men had still not been told the purpose of this vigorous testing and scrutiny, believing it was all about the possibility of test-flying a new type of high-performance aircraft.

The next step saw a delegation of Soviet doctors arriving in Prague in October 1976, led by veteran cosmonaut Vasiliy Lazarev. The delegation conducted their own examinations from 21–25 October, which finally reduced the field to just four candidates: Ladislav Klima, Oldrich Pelcak, Vladimir Remek and Michael Vondrousek. "The decision was the Russians, of course … they had the last word," said Dvorak.[3]

The four candidates arrive in Moscow. From left, Vladimir Remek, Oldrich Pelcak, Michael Vondrousek and Ladislav Klima. (Photo: Oldrich Pelcak Archives)

On 10 November, Dvorak flew to Moscow with the four finalists for two weeks of intense scrutiny and examination. A fortnight earlier, the four candidates had finally been informed of the purpose for all the testing by the Deputy Assistant Secretary of Defense, Jozef Remek (for the record, the father of Vladimir Remek). "He talked to us and he told us the top-secret information," explained Ladislav Klima. "We were not allowed to talk about [it, even] at home, but I told my wife. I trusted her."

Still more evaluations and tests, including riding a centrifuge, were carried out within Star City from nine in the morning until six in the evening. "In the evenings we talked; we were not allowed to leave Star City," reflected Klima. Alcohol was strictly forbidden. They could attend the center's movie theatre but as Oldrich Pelcak recalled, it "projected every evening the same film – one of western Kazakhstan, with Red Army soldiers fighting the bandits". They soon found other things to occupy their evenings.

In mid-November the four candidates appeared before the Soviet medical commission, where they were told that all of them were suitable candidates to fly into space. As recalled by Klima, "The Russians said, 'We were asked to test you. Now it is the decision of your government.'"[4]

The four Czech candidates returned home, and the following day it was confirmed that the two leading candidates were Remek and Pelcak, although the slightly chubby Remek had to follow a strict diet to lose around 20 pounds, in order to meet the strict weight requirements and qualify for Soyuz mass calculations. As his mother Blanka told the CTK news agency, "He loves good food and cooking, and it was a sacrifice. But thanks to his enthusiasm for sport, he managed it by playing hockey, football and volleyball. He often came home with bruised knees and elbows."[5]

After being congratulated on their selection, the two pilots were told that they would be returning to Moscow on 6 December 1976 for TsPK training, along with the chosen pair of candidates from Poland and East Germany.

TRAINING BEGINS

Oldrich Pelcak was born on 2 November 1943 in the Czechoslovakian city of Zlin, in south-eastern Moravia. At the age of seven he began his eight-year elementary education at a school in Kyjov. In 1958 he began studying at the Ugerske-Gradiste Engineering Junior College, graduating four years later in 1962. While attending high school he also took on work at the local Aero Club, where he flew gliders, later moving on to motorized airplanes. After graduating from high school he put in a tender for acceptance into the Kosice Air Force College and was accepted. He joined the Communist Party in 1964. A determined and talented student, he graduated from the college with honors in 1965 and then served as a flight engineer lieutenant with fighter squadrons in the Czechoslovakian Air Force.

In 1969, he achieved the rank of lieutenant commander and two years later qualified as a pilot 1st class. He was then promoted to the rank of captain. In 1972 he was chosen to study at the Gagarin Air Force Academy, along with Vladimir Remek. Three years later he was awarded the medal "For Service to Country" and in 1976 he defended his thesis and graduated.

Pelcak, his wife Hannah and their two sons Oldrich and Milos, were allocated an apartment inside Star City, and he recalls that one of his neighbors was Yuriy Isaulov, who had been selected to join the Soviet cosmonaut detachment six years earlier.

Soviet Cosmonaut Yuriy Fyodorovich Isaulov. (Photo: Spacefacts.de)

Isaulov was to have been selected to pair with Pelcak as one of the two teams competing for a seat on the first Interkosmos mission, but soon after he was chosen for the flight it was determined that it would not be wise to fly any all-rookie crews following the failed docking of Soyuz-25 in October 1977, especially not with a minimally trained international candidate. As a result, Isaulov was replaced by Nikolay Rukavishnikov, a veteran of the Soyuz-10 and Soyuz-16 missions. The luckless Isaulov would never journey into space. In November 1980 he was selected for the Soyuz T-3 back-up crew, but this was dissolved when its commander, Vasiliy Lazarev, failed a medical test early in 1981. With no prospect of a mission, Isaulov resigned from the cosmonaut detachment the following year.

After six months of classroom training on the theory of space flight, two potential crews were announced for the first Interkosmos mission. Vladimir Remek was partnered with Col. Aleksey Gubarev, who was commander of the Soyuz-17 mission which visited Salyut-4 in January 1975 with flight engineer Georgiy Grechko. Oldrich Pelcak joined the replacement commander Nikolay Rukavishnikov, who was a civilian. Following this, all the training was undertaken as crew units, pending an evaluation of which team was the stronger and better-trained. Each training day lasted from eight to ten hours.

Prior to his assignment to train with Gubarev, Remek had been living in a hotel room by himself. This solitary situation seems to have changed when the crew formed. The *Moscow News* reported that Remek lived with Gubarev while training. As Gubarev has pointed out: "Ever since the day we started our joint training I told him to abandon his bachelor room in the hotel and to move into my place. It was a decision made by all of us – wife, daughter, and son (also Vladimir) included."[6]

Aleksey Gubarev (left) with Vladimir Remek. (Photo: Spacefacts.de)

Beginning in October 1977, the two crews were subjected to splashdown and survival training, an exercise in which even the hardiest of stomachs were tested. As related by Pelcak, "Just once I was sick [while] practicing splashdown [procedures]. Landing in the cabin, we were on the Black Sea [with] six-foot waves and suddenly Nikolay Nikolayevich [Rukavishnikov] takes the [sick] bag and throws up. I smelt that sour smell and heard the sounds; it took a lot of overwhelming [these sensations] but after forty minutes I started to vomit too. It was an hour of torture."[7]

Oldrich Pelcak (top) with Nikolay Rukavishnikov. (Photo: Spacefacts.de)

Early in 1978 the two crews were given exercises in weightlessness aboard a special aircraft that would fly parabolic arcs over a five-hour period.

The date was 25 February 1978, and it was a day both crews had been anticipating yet dreading because the verdict was to be delivered on the flight assignments. The four men were called into the office of Lt. Gen. Vladimir Shatalov, the Commander of Cosmonaut Training, who said that following their recommendations to Prague, a decision had been reached and conveyed back to Shatalov. He then informed them that the prime crew would consist of Remek and Gubarev, with Pelcak and Rukavishnikov acting as their back-up crew.

Although he said nothing at the time, Pelcak still believes that there was some measure of political intrigue in the selection of Remek, due to his militarily influential father. However he also understands that while Gubarev had previous experience as a spacecraft commander, Rukavishnikov had only flown as the crew engineer on his two space missions.

Deputy chief of the cosmonaut training unit, Aleksey Leonov, in discussion with Oldrich Pelcak and Nikolay Rukavishnikov. (Photo: Author's collection)

A publicity photograph of the Soyuz-28 crew prior to their space mission. (Photo: Author's collection)

FIRST CZECH IN SPACE

Vladimir Remek recalls making the final journey to the Soyuz capsule with a sense of unease and even dread. "It was freezing, and there was snow on the barren steppe landscape. On the various platforms alongside the rocket were a lot of tech staff finishing up the required tasks. [They] wore long heavy furry sheepskin coats with furry hats and heavy duty gloves … and because they were working on the high platform where you really couldn't put fences, they had thick belts around them secured with heavy chains to the platform … And of course in this cold weather, the steam came up from their mouths. [As] we slowly climbed up to the top in the slow elevator … each of them, on the multiple platforms, knocked on the elevator window and showed thumbs up. I described it as if the devils escorted you to the cockpit … [because] they were hairy, furry, with red cheeks, steam coming from their mouth and those heavy, heavy chains … one hopes they are not escorting you to hell."[8]

On 2 March 1978, after an unexplained three-day launch delay, 29-year-old cosmonaut-researcher Capt. Remek finally broke the 17-year monopoly on nationalities in space held by the United States and the Soviet Union. At 15:28 GMT (UTC), Soyuz-28 lifted off from Launch Pad 1 at the Baykonur spaceport on a mission to dock with the Salyut-6 space station. Their radio call sign was *Zenit* (Zenith), and the flight was scheduled to last 7 days and 21.5 hours (plus/minus one hour). The designated duration would be the same for all the ensuing Interkosmos missions, so that no participating country could claim that it had been granted a longer time in space than any other.

The launch was attended by a delegation from Czechoslovakia that was headed by Jozef Lenart, a member of the Presidium of the Central Committee of the Communist Party of Czechoslovakia. In a political sense, the mission was also a propaganda coup for Soviet-Czechoslovakian relations, occurring on the 30th anniversary of the Soviet-backed coup d'état in 1948, when the Communist Party of Czechoslovakia assumed control over the government of that country, marking the onset of a Communist dictatorship which held power until the 'Velvet Revolution' of 1989 brought about the nation's conversion to a parliamentary republic. But that was more than a decade in the future when Soyuz-28 triumphantly took to the skies.

As Soyuz-28 flew its rendezvous in preparation for docking with the 19.8-tonne space station, the resident crew of Salyut-6, Yuriy Romanenko and Georgiy Grechko, who had arrived at the station on 11 December 1977 aboard Soyuz-26, were little more than a day from surpassing the American space record set in 1974 by Skylab astronauts Gerald Carr, Edward Gibson and William Pogue of 84 continuous days in space.

Romanenko and Grechko had previously been visited by the Soyuz-27 Soviet crew of Vladimir Dzhanibekov and Oleg Makarov. The main objective of that mission had been to exchange the Soyuz spacecraft, thereby freeing up a docking port for a Progress resupply tanker. Unlike its predecessors, the Salyut-6 station had been designed with two docking apertures; one at the forward end and another at the rear so that two visiting ships could be docked at the same time. But only the aft port had the facilities to transfer propellant to the station. The Soyuz-25 mission had had to abort after failing to dock at the front port, so the next mission had used the aft port. The Soyuz-27 mission had been improvised in order to clear the aft port and restore the configuration needed to receive for the tanker.

The Soyuz rocket prepared for launch from the Baykonur pad. (Photo: Author's collections)

Gubarev and Remek responding to questions posed before their flight. (Author's collection)

This was achieved on 16 January 1978 when Dzhanibekov and Makarov returned to Earth in the old spacecraft after spending five days aboard the orbiting station, leaving their own ship at the front port.

As the Soyuz-28 flight proceeded, the Soviet news agency TASS announced that the new mission had initiated a research program of space flights which would be followed later in the year by cosmonauts from Poland and the German Democratic Republic. The agency further reported that a group of ten cosmonaut candidates from Hungary, Cuba, Mongolia, Romania and Bulgaria would begin training that month for similar flights in the future.

TASS also said that in his first radio communication session with ground control Gubarev "reported that they had started implementing the program of preparing the Soyuz-28 ship for a rendezvous with the orbital complex Salyut-6/Soyuz-27". These pre-docking preparations included a correction in the Soyuz capsule's orbital trajectory.

The link-up with Salyut-6 took place on 3 March at 17:09 GMT, with Gubarev docking at the aft port, which the Progress tanker had vacated on 7 February. After checks to ensure the docking unit was completely airtight, Gubarev and Remek opened their hatch and made their way into the Salyut-6 station. Once there, Romanenko and Grechko gave the new arrivals a traditionally hospitable welcome to what would be their shared home and workshop for the next few days. The Soyuz-28 crew had not come empty-handed, bringing with them letters, newspapers, and magazines for their stellar hosts.

Gubarev (top) and Remek prior to insertion into their Soyuz-28 spacecraft. (Photo: Author's collection)

LIVING AND WORKING IN ORBIT

Once Gubarev and Remek had settled in, the planned work program aboard the space station began. The experiments performed included one called Oxymeter which studied the oxygen regimen in the tissue of a person while in a state of weightlessness, and another that studied the growth of chlorella algae in weightlessness. Observations were also made of the Earth, especially of glaciers and the snow mantle of certain areas. And variations in the brightness of stars were measured as they were setting behind the night horizon of the planet. As well, Remek conducted experiments on glass and other semi-conductor materials using the Splav processing furnace inside the Soyuz-28 spacecraft.

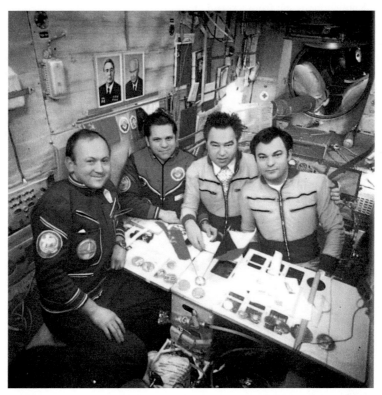

Vladimir Remek, Aleksey Gubarev, Georgiy Grechko and Yuriy Romanenko aboard the Salyut-6 space station. (Photo: Author's collection)

On 4 March, Communist Party General Secretaries Leonid Brezhnev and Gustav Husak sent a message to the crew reading: "International cooperation in space is one more proof of the fraternal relations between socialist countries and fresh evidence of the strength of socialist internationalism."

According to the flight plan, one day was set aside for leisure; a day when the cosmonauts summed up the results of their experiments, exercised, and prepared research reports. The resident crew also wrote personal letters for dispatch to Earth with the returning crew. Supplemental to this, the cosmonauts actually performed the work of postal workers by cancelling several letters with a 'Space Mail' postmark.

On 9 March, with their work program completed, Gubarev and Remek began the task of packing up their experiments and loading scientific instruments and material into the Soyuz-28 capsule. They would return carrying silver chloride capsules and copper and lead chloride samples from the Salyut-6 materials science experiments.

The following day, at 10:23 GMT, they undocked from Salyut-6's No. 2 docking port, and then prepared themselves and their craft for the return to Earth. They performed the de-orbit burn at 12:55 and discarded the now unneeded sections of the spacecraft. As Remek recalled:

"When [I was] returning with Gubarev to Earth and our cabin flew through the dense layers of the atmosphere and from a speed of almost eight kilometers per second suddenly decelerated by air friction, I had all kinds of odd feelings. Such a two-and-one-half-tonne capsule is none too big. One sits in it in a crouched position, there is a wall close by, on the far side of which is heat of over one thousand degrees Celsius. It is all thoroughly thought out and calculated but something might still happen; a hole might be burned in the bottom, the cabin may begin to break up and melt down … The thought comes inevitably to one's mind that this may be the end."[9]

Despite Remek's apprehension, the re-entry went as planned. At 13:45 they made a soft landing some 90 miles north of Arkalyk in northern Kazakhstan. The mission had lasted a total of 7 days, 22 hours and 16.5 minutes and from all accounts it was a complete success, despite a lack of precise mission details. The condition of both men was later deemed to be excellent.

The Soyuz-28 cosmonauts are interviewed following their flight. (Photo: CTK)

As commented by Gen. Aleksey Leonov, then deputy chief of the cosmonaut training unit, the Soyuz-28 flight was "an excellent beginning to a promising task for the countries of the socialist community. There will soon be flights by spacemen from Poland and the German Democratic Republic, and in subsequent years of representatives of other socialist countries."[10]

One amusing story followed, when stories began to circulate of Remek returning to the ground with red hands, and Western analysts expressed concerns about possible circulatory problems. However it was all a concocted story. As James Oberg explained in his seminal book *Red Star in Orbit*, the story stated that when doctors examined Remek after his flight and pointed out his bright red hands, the cosmonaut had looked at his hands and laughed. "Oh, that's easy," he replied. "On Salyut, whenever I reached for a switch or dial or something, the Russians shouted, 'Don't touch that!' and slapped my hands."[11]

From that time on, many of the guest cosmonauts had to put up with comments about red hands.

HOME TO ADULATION

After returning to Earth, Remek received the title of Hero of the Soviet Union and the Order of Lenin; the honorary degree of Pilot-Cosmonaut of the Czechoslovak Socialist Republic and the title of Hero of Czechoslovakia with the Order of Klement Gottwald; the Gold Medal of the Czechoslovakian Academy of Sciences; the Medal for Service to the Motherland, and various other awards and medals.

Early in May, thousands of people turned out in Prague to greet Remek and Pelcak, plus Soviet cosmonauts Aleksey Gubarev, Yuriy Romanenko and Georgiy Grechko, all of whom had been involved in the first Interkosmos joint flight. They were accompanied by Vladimir Shatalov, then in charge of cosmonaut training.

In March 1979, a year after completing his Soyuz-28 mission, Remek married a popular 34-year-old Prague film and television actress named Hana Davidova, the daughter of former Czechoslovakian foreign minister Vaclav Davidov. They had a daughter named Anna in 1980 who would also go on to become an actress. Unfortunately the marriage would later end in divorce. Hana died in Prague on 23 May 2006, aged 63.

In late 1979, Remek became Deputy Director of the Military Research Institute in Prague; a position that he held for six years. In 1985 he joined the Defense Office of Czechoslovakia, but as a national icon he was no longer permitted to fly aircraft, a restriction that he had great difficulty in accepting. Finally, in January 1986, he returned to flight operations as Deputy Commander of the Flight Division in Czaslau, 40 miles out of Prague, and was at last able to resume flying aircraft. As he reflected in the 1990 book, *Cosmic Returns*:

"I used to have the feeling sometimes that no one wanted anything else from me, or ever would, but to act as the country's representative who would forever go on talking about his experience in space during the eight days of my stay there. I willingly handed my spacesuit over to the Aviation and Cosmonautics Museum but I have always been against becoming a walking museum exhibit. As a pilot I was delighted when I finally received permission to go up again and fly. When during my travels all over the country our airmen used to take me in their transport plane I could not resist going up to the cockpit. They noticed how keenly I watched their hands clutching the control lever. We airmen know each other in most cases. One of my friends once told me: 'I see that you'd like to get a grip on that handle once again. Come on then! Take your seat for a while.' I was pleased to take his place and taste to the full the joy of moving the controls and feeling the subordination of the machine.

"Flying is, of course, a pilot's passion. We yearn for it, probably more so than the smoke addict for his cigarette. I am delighted to be able to put on quite regularly my pilot's overalls and helmet, and climb into the cabin of a jet aircraft. I would, of course, be happy to sit once again in the seat of a piloted spacecraft, but I am quite satisfied to see someone else enjoying this unique experience."

In the fall of 1986 Remek attended the Military Academy of General Staff. Graduating in mid-1988, he became Deputy Commander of the Air Defense Division in Moravia. As he said of that period of his post-flight life and the sometimes unwanted ramifications of fame:

"Today, interest in talks with a cosmonaut has somewhat slackened off as I have already told all I know, and I could hardly add anything to what I've already said at least a thousand times. I am therefore concentrating now in the first place on the fulfillment of my duties as a member of the Air Force of the Czechoslovak People's Army. I also have a little more time left for my private affairs. People are, of course, also interested in that part of a cosmonaut's life on Earth.

"Popularity has its reverse side as well. More than once I wanted to enter a shop and buy something unrecognized – like anyone else. Or board a tram, travel to Wenceslas Square and buy a sausage with mustard at a kiosk, have a glass of beer and go by Metro to the other side of town. Or on a Sunday, stroll down the street or round the park without feeling fixed eyes boring straight through me or hearing behind my back: 'Isn't that Remek?' How nice it would be not causing any excitement, curiosity and the rioting of autograph hunters … And when waiting for a late night tramcar, not to inspire drunkards or rowdies. And also not to gratuitously stir up envious people who gleefully wish upon you every slip or mishap and sometimes even do their best to bring them about."[12]

From 1990, now with the rank of colonel, Remek served as the Director of the Military Air and Space Museum in Prague-Kbely, but in June 1995 he loosened his ties with the military on a bitter note, requesting permission to leave the Czech Army after disagreements with the Ministry of Defense. An organizational restructuring earlier that year had resulted in his department being disbanded. As an alternative he was offered the position of curator of the museum's collection of around 150 space artifacts – many of them relating to his space flight and the Soyuz-28 capsule. Remek said he felt uncomfortable with the idea of being subordinated to the position of keeper of his own memorabilia, and in a sense becoming an additional artifact in the collection. He found it an untenable situation, forcing his hand.

"I've been trying to avoid that position all my life," he stated at the time. "I won't deny that, for some people, it would be the most acceptable solution to make me into another exhibit in the museum. I wouldn't go so far as to say that they would stuff me." He was incensed at what was being suggested, saying publicly and frankly that the Army had dealt with him on an "unprofessional level" and the position he had been offered – even on the same salary – was an insult to his status, experience and specialized education. "It's a question of position and work satisfaction," he complained. "I'm capable of more."

By way of official response, Peter Fuzak, a press officer for the Ministry of Defense, said he couldn't comment on the alternative position offered to Remek or the allegations that the famed cosmonaut had received bad treatment from the military. He emphasized that Remek had requested to leave on his own initiative, saying in conclusion, "He asked to leave, and we accepted."

Similarly, Boris Domin, deputy director of the historical institute of the Czech Army, went on the record saying he didn't believe that Remek had any grounds for claiming he had been badly treated. "Since 1990, he was given every opportunity. He was very often abroad with his wife. He had the maximum support. He was abroad more often than any other colonel in the Army."

Remek did add that despite his rigid stance on the matter, he had no regrets about either his military career or his time with the museum. As for the future, he declared, "I think I'll find a way to put to use my life experience, contacts [and] knowledge of the territory of the Soviet Union. It's more useful to be alive than to be an archive [exhibit]."[13]

Having carried through with his resignation from the Czech Army, Remek became a sales representative for CZ Strakonice in the Russian Federation and was CEO of the joint venture CZ-Turbo-GAZ in Nizhny Novgorod, 250 miles east of Moscow.

A TIME OF DANGER

On Sunday, 28 October 2001, Vladimir Remek and visiting U.S. astronaut Gene Cernan survived the crash of a Russian-built helicopter which unexpectedly ran out of fuel while airborne.

The two space explorers were on their way to visit Bernatice, the hometown of Cernan's grandfather Jozef Cihlar in the Czech Republic, when the engine of the Mi-8S helicopter suddenly cut out, causing the craft to fall around 500 feet to the ground near Milevsko, 60 miles south of Prague. The pilots skillfully managed to accomplish a hard landing without any loss of life, although there were several injuries. Remek was taken to hospital in the town of Pisek for observation and Cernan was transported to the Central Army Hospital in Prague with suspected broken ribs.

Gene Cernan, John Blaha, Vladimir Remek and Oldrich Pelcak at a press conference, with a translator seated at left. (Photo: Pagi)

Cernan and fellow astronaut John Blaha had arrived in the country two days earlier at the invitation of Jiri Sedivy, chief of staff to the Czech Ministry of Defense. Blaha, whose Czech grandfather Antonin Blaha came from Heralac in Bohemia, was not with Cernan at the time of the crash. It was actually the second time Cernan had survived a helicopter crash.

On 23 January 1971 he was flying a NASA helicopter which went nose-first into the Banana River, near the Kennedy Space Center.

Two days after being admitted to the Prague hospital for precautionary observation, and receiving treatment for a slight injury to his left foot, Cernan was released. Of the twelve passengers who were aboard the helicopter, he was the first to leave the hospital. Among those who remained bedridden for observation was Vladimir Remek.

The Mi-8S helicopter following the crash south of Prague. (Photo: Virtulnik.cz)

While convalescing in hospital, Cernan had told Czech Radio how the actions of the pilot and crew almost certainly saved the lives of the passengers on board. "We came down, the engines quit. We don't know why. But I can tell you without knowing anything else that the pilots and the crew reacted to avert an even bigger accident." He particularly praised the quick reactions of military police officer Jaroslav Selong, who smashed one of the windows and began dragging the passengers out of the helicopter. Cernan also reserved special praise for Remek, who went back to the craft to pull out more survivors.

"He said 'Gene, we got to get out of here, got to get of here,' and I could walk on one leg, so he pushed me through a window, it was only about so big. Someone pulled me out the other side and then he got out and together we hobbled across the field. It was damp, it was raining. And I just sort of fell to the ground, and he – God bless him – got up and went back to help more people. Another hero of the moment."

He was asked how he felt. "Right now? How do I feel right now? I feel like someone sealed me up in a big wine barrel and threw me over Niagara Falls. I'm okay. I'm fine, and when I leave here I will leave here knowing I had the best imaginable care possible. When I leave here Wednesday morning I'll be hurting – I'll have some sore bones, but there will have been nothing more that could have been done for me no matter where I had been."[14]

The cause of the incident was eventually traced to a red/green indicator light on a fuel transfer gauge that had somehow been installed upside down. From his back seat position, Cernan had actually noticed the light on this helicopter was different to that on the same model they had flown in the previous day.

LATER ACCOMPLISHMENTS

From 2002 to 2004 Remek was the commercial counselor and head of the commercial and economic section of the Embassy of the Czech Republic in Moscow. In June 2004 he was elected deputy to the European Parliament as an independent candidate for the Communist Party of Bohemia and Moravia and, being second on the list, was elected into the European Parliament. In 2009 he defended his mandate for a further five years.

Once celebrated as a hero throughout the Eastern Bloc, Remek's service as a political figure caused some controversy in his homeland. His critics, who had once lauded him for his epic space mission, pressed the point that his flight had primarily served a political and propaganda purpose. They stated that his flight into space was apparently organized in 1978 to commemorate the 30th anniversary of the Communist coup in the former Czechoslovakia and to soften anti-Moscow feelings following the Soviet invasion of 1968. However, Remek vigorously denied this.

"I don't think so. If the Soviet leaders had any problems at that time it wasn't a sense of guilt for entering Czechoslovakia. It could have been partly political, but what was really important was that we were among the strongest partners in the Interkosmos program, and our people were also on the U.N. space committee. And maybe we weren't the worst among those who prepared for the flight."[15]

The same critics said Remek was an ideal candidate for the job of carrying the Czech flag into space. Born in the year of the Communist putsch, he symbolized the unity of his former country thanks to an influential Slovak father and Czech mother – claims which a defensive Remek dismisses.

"Flights into space, especially the first cosmonauts and trips to the Moon, are always the subject of a kind of political propaganda, under Soviets and Americans alike," he stressed. "But I can tell you that in the first few hours as we orbited the Earth, we were picking up reports on our receiver and we knew that people around the world were talking about us … Our space flight was used as propaganda, but all such propaganda has to be based on work successfully carried out, which our flight was." But he also allowed himself a humorous aside, saying, "It is easier to fly into space than to make European legislation."[16]

Following his two terms in the European Parliament, Remek became Czech ambassador to Russia, submitting his credentials to President Vladimir Putin in the Kremlin on 16 January 2014. From his viewpoint, he has always remained firmly loyal to his Communist roots and views, and steadfastly promotes the ideology and party that educated him and gave him the opportunity to be selected to fly into space. As well, as an iconic figure in Czech culture, he continues to enjoy a relative amount of success in the now democratic Czech Republic.

Vladimír Remek married twice and has two daughters – Anna from his marriage to Hana Davidova, and another girl from his second marriage to Jana, who was named Jana after her mother.

On the 30th anniversary of his space mission, Remek received the great honor of having his name applied to Asteroid No. 2552, discovered on 24 September 1978 by the Czech Klet Observatory.

Remek and his former Soyuz-28 commander Aleksey Gubarev in 2013. (Photo: Jakub Dospiva/CTK)

For his part, after returning home, Oldrich Pelcak held several command positions within the Czech Air Force and then operated for twelve years as a test pilot with the Department of Aerospace Air Force in Prague-Kbely. He remained with the Air Force until 1998, a total of 33 years in the military service of his country, and retired holding the rank of colonel. From 1999 until 2013 he was employed as an insurance agent at the Generali Insurance Company based in Prague and is now retired and a pensioner.

A recent photo of Oldrich Pelcak. (Photo: Zdroj/CTK)

Despite feeling that he lost out on flying the Soyuz-28 mission owing to the influence of Vladimir Remek's father, Pelcak is still willing to participate in meetings in order to help to popularize cosmonautics.

He and his wife Hannah now reside in Ujezde nad Lest, in eastern Prague. In addition to their two sons and two daughters they now have two grandchildren, Martin and Oldrich. As with Remek, astronomers named an asteroid in Pelcak's honor on the 30th anniversary of the Soyuz-28 mission, namely Asteroid No. 6149 which was discovered on 25 September 1979.

The scorched Soyuz-28 capsule is now on permanent display in the Air Force Museum of Flight at Prague's original airport in the suburb of Kbely, five miles north-east of Prague's city center. Inside the cramped interior of the spacecraft is a single space-suited mannequin representing Vladimir Remek, the first Czech citizen to fly into space.

SOYUZ-28 MAJOR EXPERIMENTS SUMMARY

Khlorella-1 (Chlorella, or green freshwater algae): A study of the effects of microgravity on the growth of unicellular algae, which served as a model for fast-growing organisms. All the samples were carried in soldered ampoules stored in four containers with a nutrient medium. The aim of the experiment was to learn more about the future use of algae in air and water recycling systems aboard spacecraft. It was determined that microgravity does not affect the speed at which algae populations grow.

Morova-Splav and Morava-Kristall (Morova is a river in Czechoslovakia; 'splav' means alloy; 'kristall' means crystal): A series of materials-processing experiments with the aim of producing materials in space that cannot be efficiently obtained on Earth. The experiments, prepared jointly by the Solid Body Physics Institute of the Czechoslovakian Academy of Sciences and the Institute of Space Research in Moscow, studied the solidification of melted crystalline and glassy materials and the growth of crystals from the gas phase.

Kislorod (Oxygen): An experiment to study any changes in the supply of oxygen to various parts of human tissue during flight and to see how human tissue uses oxygen in microgravity. The experiment used a portable Czech-built device called an Oximeter which consisted of an array of special sensors.

Oprosnik (Questionnaire): Twice during the flight the cosmonauts filled out a questionnaire on their medical and psychological condition. The questions related to such things as their consumption of food and water, their sleep patterns, their eyesight, their senses of smell and hearing, and their aesthetic requirements. These were rated on a scale of five. Developed by Soviet, Czech and Polish specialists, this experiment was designed to help determine how people adapt to the unusual conditions in space and to further improve means of living and working in confined spaces.

Teploobmen-2 (Heat Exchange 2): An experiment designed to study the cooling effects of the space station's atmosphere. The cooling of objects that produce heat greatly changes in weightlessness because one of the main elements of heat exchange – namely heat emission through natural convection – is missing.

Ekstinktsia (Extinction): The crew visually observed how the brightness of stars changed as they disappeared behind the horizon. Earlier Soviet and American crews had noticed that stars begin to fade whilst still about 100 km from the horizon. At the time, no satisfactory explanation had been found for this phenomenon. The cosmonauts' task was to determine changes in the magnitude, color, and scintillation of stars as they approached the horizon. Their observations would help the Astronomical Institute of the Czechoslovak Academy of Sciences in Ondrejov to develop a photometer to undertake similar observations on future missions.

REFERENCES

1. *East Europe Report, Scientific Affairs, No. 751*, Joint Publications Research service, Arlington, VA, 10 August 1982
2. Tereza Pultarova interview for *3 Eyes* space community magazine article, "Vladimir Remek – The First Non American, Non Soviet in Space," issue 20 December 2011
3. Petr Fencl, *Infofila* online magazine, article "Kosmos 3/2003: Než vzlétl náš kosmonaut," dated 12 November 2003. Website at: *http://www.infofila.cz/kosmos-3-2003-nez-vzletl-nas-kosmonaut-r-2-c-637*
4. *Ibid*
5. *Sydney Morning Herald* newspaper, article "Dieted: now weightless," issue 3 March 1978
6. Rex Hall papers, retrieved by Bert Vis, September 2014
7. Petr Fenel, *Infofila* online magazine, article "Kosmos 3/2003: Než vzlétl náš kosmonaut," dated 12 November 2003. Website at: *http://www.infofila.cz/kosmos-3-2003-nez-vzletl-nas-kosmonaut-r-2-c-637*
8. Sebastian Duthy and Veronika Lukasova, "The New Space Race," *Zen* magazine, issue 9 May 2011
9. Vladimir Remek, "Cosmonaut on Earth," extract from *Cosmic Returns*, Smena Publishing House, Moscow, 1990
10. *Soviet Weekly* newspaper, "Intercosmos space flight was success," issue 18 March 1978, pg. 4
11. James Oberg, *Red Star in Orbit: The Inside Story of Soviet Failures and Triumphs in Space*, Random House, New York, 1981
12. Vladimir Remek, "Cosmonaut on Earth," extract from *Cosmic Returns*, Smena Publishing House, Moscow, 1990
13. Emma McClune, *The Prague Post* newspaper, article, "Czech Cosmonaut Unhappy About Plans to Make Him a Museum Piece," issue 14 June 1995
14. Rob Cameron, *Radio Prague* transcript, "Ex-astronaut praises crew, fellow passengers after helicopter crash," 30 October 2001
15. Ian Willoughby, "Vladimir Remek remembers his role in the Space Race," Czech Radio article, 7 March 2008
16. Radek Honzac, article "European Parliament celebrates its space member" from *Europolitics* magazine, No. 3499, 31 March 2008, pg. 24

3

From Poland to Salyut-6

Had circumstances been ever so slightly different, future Polish cosmonaut Miroslaw Hermaszewski would have died in infancy as the result of a terrible slaughter in his home village of Lipniki. The massacre, in which members of the Hermaszewski family became innocent victims in March 1943, was carried out by nationalists of the Ukrainian Insurgent Army (UPA) in the Nazi-occupied regions of Volhynia and Eastern Galicia, which had been divided between Poland and the Soviet Union early in World War Two. The killings were directly linked with the policies of the Stepan Bandera faction of the UPA, whose stated goal was the purging of all non-Ukrainians from the future Ukrainian state. Not limiting their activities to the purging of Polish civilians, the UPA also wanted to erase any trace of sustained Polish presence in the area through an 'ethnic cleansing' operation.

Close to 60,000 innocents were callously slaughtered in Volhynia and up to 40,000 in Eastern Galicia. Most of the victims were women and children.

A unit of the UPA on the march. (Photographer not known)

© Springer International Publishing Switzerland 2016
C. Burgess, B. Vis, *Interkosmos*, Springer Praxis Books, DOI 10.1007/978-3-319-24163-0_3

TINY SURVIVOR

Miroslaw (Mirko) Hermaszewski came into the world as the son of peasant small-holder Roman Hermaszewski and his wife Kamila on 15 September 1941. He would be the last born of seven children, with four sisters and two brothers: Aline, Wladyslaw, Sabina, Anna, Teresa, and Boguslaw. The family had moved to the village of Lipniki in south-western Poland at the beginning of 1943 in the hope of escaping from the terrors being inflicted on non-Ukrainians by the UPA.

In Lipniki, the family worked a seven-acre farm located on the outer edge of the former Radziwill estate. At that time Lipniki was filled with similar refugees, and Roman became one of the leaders of the defense of their village. "We had lived in harmony with the Ukrainians," Hermaszewski later declared. "Especially my uncles and grandfather, who said that they were our brothers, and that there was nothing to fear."[1]

On the evening of 26 March 1943, when Miroslaw was just 18 months old, a vicious assault was made on Lipniki by a unit of the UPA. These savage bandits, armed with pitch-forks, axes, knives, and a number of guns rushed to the village and methodically began cutting down the terrified inhabitants. Indescribable panic ensued, during which 182 people were brutally slaughtered, including 18 members of the Hermaszewski family. Those killed were mainly women, old men, and more than 50 children of 1 to 14 years of age, many from families that had lived and toiled in the village for centuries.

When the insurgents discovered that a local Polish man named Jakub Warumzer had been sheltering in Roman Hermaszewski's small house, they burned the house to the ground, beheaded Warumzer, and bayoneted Miroslaw's grandfather in the head seven times until he was dead.[2]

According to Hermaszewski, his terrified mother Kamila grabbed him and ran through a snow-covered field heading for the surrounding forest with her baby son on her back, hotly pursued by a member of the UPA armed with a gun. Desperate cries for help and pleas for mercy came from all around as the butchery continued. The insurgent fired several shots at close range and Hermaszewski's mother fell to the ground, wounded in the temple and ear. Believing she was dead the man headed back into the burning village. Fortunately for both Miroslaw and his mother he did not cry, which in all likelihood saved his life. "It was a miracle," Hermaszewski recalled. Kamila awoke several hours later. Covered in blood, concussed, groggy, and barely conscious she staggered six kilometers to a village where friends lived, and found refuge with them.

The next morning, by which time the insurgents had left the area, Roman and his 15-year-old son Wladyslaw went on a desperate search for their little Mirko. Not knowing where to start, they began to search a field strewn with dozens of corpses. Suddenly Wladyslaw found the bundled child, but his baby brother was showing no signs of life. They rubbed his hands and torso and the little boy opened his eyes momentarily. He was rushed back to the friend's house and warmed in front of a fireplace. Soon little Mirko stirred, looked up, and began to cry.[3]

Another tragedy would strike the family five months later when Roman Hermaszewski was gunned down by fascists in a corn field during fighting on 28 August 1943. Hit in a lung, he died a few days later. Kamila found refuge for herself and her seven children in a rectory in Berezne, where Father Mieczyslaw Rossowski took care of them until the war ended.

In June 1945, the family relocated to the western territories, settling in Wolowo, near Wroclaw, where Miroslaw graduated from primary and secondary schools. In his childhood he developed a fascination for aircraft and aviation. He recalled one occasion when the train that he was on had stopped at a station and he watched a small biplane landing at an adjoining airfield, bouncing a couple of times on the grass before heading off to a hangar. "This scene completely engulfed me," he told an interviewer. "I was fascinated by it."

In his youth Hermaszewski, assisted here by his sister Teresa, would design and build model airplanes. (Photo: M. Hermaszewski Archives)

While he was still attending primary school, Hermaszewski began to create rudimentary models of aircraft and rockets, and he eagerly read magazines about aviation. His interest was further fueled when his elder brother Wladyslaw enrolled at the Deblin Air School in 1948 and began to appear in the family home dressed in his air force uniform while on leave. His other brother Boguslaw also joined the air force. Miroslaw was soon tinkering with and building more sophisticated model planes at aviation model shop classes in the nearby town of Brzeg Dolny hosted by the League of Soldiers' Friends. At the age of 16 he began basic glider training at the Wroclaw Aeroclub airport and was granted his glider pilot's license a year later.

In 1961, after completing his general studies secondary education, Hermaszewski also enrolled in the Polish Air Force Officers Flying School in Deblin and, as did many others, joined the Communist Party. On graduating as a second lieutenant at the top of his class in November 1964, he first operated as a MiG-15 fighter pilot and then served as a senior pilot and wing commander flying MiG-17 and MiG-21 fighters in the Polish Air Defense Force until 1969. He next attended a three-year course in the aviation department of the Karol Swierczewski General Staff Academy's Faculty of Science in Warsaw. Meanwhile, in 1965 he had married Emilia, who was an employee of the Polish airline Lot, and their son, named Miroslaw after his father, was born in 1966. A daughter, also named Emilia, would be born in 1974. After graduating from the academy, Hermaszewski was promoted to the rank of captain and assigned as the commander of the 11th Fighter Regiment, flying MiG-21 fighters. After holding several command posts in different Air Defense Force units, Hermaszewski was promoted to major on 6 January 1975 and three months later he assumed command of a fighter plane regiment. In April the following year he was the commander of a regiment of flight instructors when the selection process began to find the first Polish cosmonaut for the Interkosmos program. By this time he had accumulated 1,480 hours flight time, of which 550 hours were accrued in supersonic airplanes.

SPACE FLIGHT COMPETITORS

It is not clear exactly when Poland initiated the selection of that country's Interkosmos candidates. According to Soviet sources, the selection began simultaneously in Poland, Czechoslovakia and East Germany immediately following the signing of an agreement with the nine socialist countries in Moscow in September 1976. However later records reveal that the long and complicated selection process – at least in Poland – began as far back as April 1976.

All activities relating to the recruitment of candidates were initially the responsibility of the Polish Aviation Commission acting through the Military Institute of Aviation Medicine and Psychology in Warsaw, and were based on university education, flying experience and acromedical documentation. In much the same way as NASA's Mercury astronauts were selected, only military pilots with jet experience would be considered. This qualification greatly simplified the first phase of the selection because the pilots had to undergo annual physical check-ups. These records were made available in absentia for the screening of the 300 potential candidates. At the conclusion of Stage I of the selection process, 71 potential candidates were identified.

Stage II of the operation was conducted by the Polish Ministry of Defense at the end of August, and this government office narrowed the field down to just 17 candidates,[4] namely:

- Capt. Marian Boleski
- Maj. Andrzej Bugala
- Capt. Boleslaw Danczak
- Maj. Janusz Dorozynski
- Maj. Kazimierz Fiolek
- Maj. Henryk Halka
- Capt. Jozef Hajduk
- Maj. Miroslaw Hermaszewski
- Lt. Col. Zenon Jankowski
- Capt. Mieczyslaw Kafel
- Maj. Szymon Kowal
- Lt. Tadeusz Kuziora
- Maj. Kazimierz Lewenko

- Capt. Czeslaw Pazur
- Capt. Bogdan Rojek
- Maj. Jan Waliszkiewicz
- Capt. Henryk Wrzesien.

When the purpose of the exercise was revealed to the candidates, Capt. Danczak asked to be excused from further participation for personal reason.

The second stage of the selection process consisted of two parts. First, on 15 September, the 16 remaining candidates were subjected to a grueling fitness test at the Military Training and Conditioning Center of the Polish Air Force, situated in the lakeside town of Sniardwy, Mragow, which was located some 3,100 feet above sea level. In the second part of the tests, the candidates were assigned to another military training center in Gronik, near the town of Zakopane in the Tatra mountains (known as the 'Polish Alps'), where they went on extensive hikes to assess their physiological and physical capabilities at high altitude.

The next step involved a two-week period of specialized training and additional medical examinations at the Military Institute of Aviation Medicine. This included flight simulator training sessions and training in various fields of cosmonautics. In addition to being tested for their tolerance to hypoxia in high-altitude simulations and their spatial orientation, they had to endure crushing g-forces in centrifuge tests.

Based on the results from this phase of the selection process, the list was reduced to five candidates: Andrzej Bugala, Henryk Halka, Miroslaw Hermaszewski, Zenon Jankowski and Tadeusz Kuziora.

Candidates from left: Andrzej Bugala, Zenon Jankowski, Miroslaw Hermaszewski, Tadeusz Kuziora and Henryk Halka. (Photo: WAF)

The five remaining candidates undergoing high-altitude testing in Warsaw. From left, the pressure-suited candidates are Andrzej Bugala, Henryk Halka, Zenon Jankowski, Tadeusz Kuziora and Miroslaw Hermaszewski. (Photo: Maceem Stolovskim)

Hermaszewski in his pressure suit for the high-altitude tests. (Photo: M. Hermaszewski Archives)

The third stage involved the participation of a Soviet medical team visiting Warsaw on 29 October 1976 to conduct their own tests, accompanied by veteran physician-cosmonaut Col. Vasiliy Lazarev. They determined that Andrzej Bugala be eliminated from the squad, as his body did not fit the anthropometric requirements; when seated he was around four inches too tall for the Soyuz spacecraft.

Maj. Andrjej Bugala was eliminated after tests showed he was too tall to fit into the Soyuz spacecraft. (Photo: Spacefacts.de)

After reviewing test results and preparations, and after conducting still more tests, the four remaining candidates were then sent to the Gagarin Cosmonaut Training Center, which was near Moscow, where they undertook an intensive preparatory course over a two-week period from 10 November, studying astrophysics and astronavigation, space medicine and biology, physics and geology, chemistry and meteorology, photography, and many other specialized technologies. In all, over 40 subjects. Along with this there were the principles of building and servicing equipment for space flight and, finally, simulated flight exercises conducted in mock-ups of Soyuz and Salyut spacecraft. All four candidates were certified as meeting the requirements, but a final judgment had to made that would bring their number down to just two.

In 1978, Col. Romuald Bloszczynski of the Polish Military Institute of Aviation Medicine was asked about the criteria used for selecting the cosmonaut candidates, and their requisite psycho-physical characteristics. Although he offered a good insight, he obviously gave some obfuscating figures to disguise certain facts about the process.

"We had selected a large group of scores of pilots from among military supersonic aircraft personnel. All those in that group have had a great deal of professional experience and were in excellent health. After preliminary tests at the Military Institute of Aviation Medicine, we reduced the group to about twenty persons and assigned them to aviation conditioning centers in Mazury and in Zakopane for two months of training, the purpose of this being to equalize the pilots' performance characteristics so as to establish a uniform comparative basis for the third selection stage.

"After training in camps where medical and psychological tests occurred in parallel, there were only twelve candidates left. Those were subjected to a cycle of all-encompassing tests sequentially administered by teams of specialists. From the final discussions and composite analyses there evolved a points-based list and four pilots were selected for additional testing in the Soviet Union. The results of these tests were favorable. All four were recognized as being completely fit for space travel. Our government decided to assign two of them to the basic training of cosmonauts, as official representatives of Poland."

Bloszczynski was then asked which psychological characteristics of the candidates were given special attention.

"The tests were all-encompassing, but actually three questions were most crucial. I have in mind the capability of the circulation system, the effectiveness of the balancing organs, and general resistance to stress. These are the factors of extra importance to be considered under space flight conditions.

"We subjected the candidates to tests such as, for example, centrifuges and variable loads in a so-called cyclergometer. Throughout these tests we recorded physiological parameters such as ventilation of the lungs, pulse, blood pressure, and electrocardiograms. On spinning chairs and other devices, we tested the behavior of their balancing organs under a variety of perturbations. In low-pressure chambers we determined the so-called system reserve capacity for conditions of heavy oxygen depletion. By the way, these are only examples of the tests; an exhaustive description would fill a few hundred pages."

And finally, what were the IQ scores of the chosen Polish cosmonauts?

"Excellent. We applied, amongst others, the Wexler norm adapted to the overall Polish population. The results fell within the 130 to 135 range, the norm for an adult person being 100. Scores like those of the cosmonauts are attained by barely a few percent of society."[5]

The selection of the two finalists was made on 27 November 1976, and those approved for cosmonaut training were Miroslaw Hermaszewski and Zenon Jankowski. Together with the pairs of candidates from Czechoslovakia and East Germany, they would now undertake two months of preparation theory, a crash course in learning the Russian language, and becoming intimately familiar with the construction, systems, and operation of the Soyuz spacecraft and Salyut space station.

In regard to the other three Polish candidates:

- *Lt. Col. Henryk Hałka* was born on 18 February 1941 in Olchowiec. He graduated from the Academy of the General Staff of the Polish Army in Rembertów in 1979, was appointed commander of a fighter regiment in Goleniów/Szczecin in 1980, and died on 12 December 1980 when his MiG-21 crashed in bad weather near the air base in Goleniów.

- *Lt. Tadeusz Kuziora* was born on 3 March 1949 in Ternow. He graduated from Deblin officers' school in 1971, graduated from the Academy of the Polish Army General Staff in 1978, was appointed commander of the 7th Cavalry Regiment bomber unit in 1984, was promoted to the rank of brigadier general in November 1994, and became commander of the Deblin Air Force Military School in 2002.
- *Maj. Andrzej Bugala* was born on 5 February 1940 in Bartkowice. He made his first parachute jump in 1956, graduated from High Officers' summer school in 1962, was sent to study at the Gagarin Air Force Academy in Moscow in 1970, was the Polish Air Force Chief of the Krakow military district, retired with the rank of colonel, and died in Krakow on 10 July 2013.

SECOND CANDIDATE

By comparison with Miroslaw Hermaszewski's turbulent early life, Zenon Jankowski had enjoyed a relatively normal upbringing. He was born in Poznan, Poland, on 22 November 1937, where he received his primary and secondary education. On graduating in 1956 from the Marcin Kasprzak general education school (named for a famed Polish revolutionary), he enrolled in an Air Force Officers' School in the city of Radom, south of Warsaw. The next year he continued his aviation studies at the Polish Air Force Academy in Deblin, in eastern Poland, achieving the status of military pilot 3rd class in 1958 and graduating as a lieutenant on 13 March 1960. After serving as an instructor in various aviation units, he flew the MiG-15 as a military pilot 2nd class from 1961. Starting in 1962, now a military pilot 1st class, he flew the MiG-17 with different fighter-bomber units. It was during this time that Jankowski married his girlfriend Aniele, and their daughter Katerina (Kasya), who would be their only child, was born in 1964.

In 1966 he entered the Karol Swierczewski General Staff Academy of the Polish Army, from which he graduated in 1969. Then he served as a flight commander and later squadron commander. In 1976, now a lieutenant colonel, he was appointed deputy commander of an air force regiment and also served as a commander of high-ranking officers in the aviation school in Deblin. That year he also participated in the 'Tarcza 76' Warsaw Pact exercises, and professionally commanded a squad of variable-wing airplanes. Shortly after his return from the exercises, he was ordered to report to the Military Institute of Aviation Medicine. He and several dozen other air force officers were then subjected to a program of tests and preparations, the aim of which was to select Polish candidates for the Interkosmos program. Jankowski was ultimately selected along with Maj. Miroslaw Hermaszewski, and the two men began their training at the Gagarin Cosmonaut Training Center on 4 December 1976.

The training program for the Interkosmos candidates was both intensive and demanding. The first stage centered on theoretical studies but was combined with flight experience in jet airplanes, physical exercises, simulation of zero-g in aircraft which flew high ballistic arcs, and elementary recovery procedures including the possibilities of splashing down on water and retrieval by helicopters from dense forests, mountains, and other difficult terrain. The second stage comprised of mastering the Soyuz spacecraft, together with the specific flight program for the planned Soviet-Polish mission aboard the Salyut-6 space station.[6]

Lt. Col. Zenon Jankowski. (Photo: Spacefacts.de)

CHOSEN TO FLY

In July 1977 Hermaszewski and Jankowski would each be paired with their respective Soviet cosmonaut commander. Hermaszewski was assigned to Pyotr Klimuk, who had flown both Soyuz-13 and Soyuz-18, while Jankowski trained with Valeriy Kubasov, a veteran of Soyuz-6 and Soyuz-19. The paired crew training began in August, but Hermaszewski's training was interrupted when he was hospitalized with what Russian doctors informed him was tonsillitis and had to undergo a minor operation. He recovered quickly and was soon back training with Klimuk.

On their training, Jankowski observed that "in our case it took almost two years. For the first four months we received the basic theoretical background pertaining to the dynamics of flight, astronomy, propulsion systems, etc. At this time we were also introduced to difficult technical language. For the next several months we combined theory with practice dealing with matters of rescue, high energy rides on centrifuges, astronavigation and such. The third stage lasted one year and included training in specific crew functions. Here we implemented the timing of the flight program."[7]

Only one of the teams could fly. Klimuk and Hermaszewski would be confirmed as the prime crew, although their names would not be officially announced until 25 June, just two days before the flight. This was in line with the policy surrounding Interkosmos crewing; the choice of who would fly was always made at the time of the crew formation, several months before the official naming announcement. But the back-up crew continued to train hard, just in case they had to suddenly switch to the prime crew position in the event of injuries, illness or any other critical reason.

The back-up crew of Kubasov and Jankowski during training. (Photo: Author's collection)

Prior to the flight, Col. Stanislaw Baranski, commandant of the Military Institute of Aviation Medicine addressed the question of what Hermaszewski would be doing aboard the space station. A professor and doctor, Baranski was a prominent Polish expert in the field of aerospace medicine, and also a coordinator of preparations for the flight of Soyuz-30. These were his comments:

We have prepared for our cosmonaut a research and test program which is rather broad in scope and scientifically important. It covers as many as 12 subject matters, including nine in the field of medicine, physiology, and psychology, two in the field of physics and space technology, and one in the field of Earth science or, more precisely in photography of Polish territory for various uses in science and the national economy.

The prime crew of Klimuk and Hermaszewski in training. (Photo: Author's collection)

The most important experiment in the field of space medicine, now performed by us jointly with the Soviet Union, includes testing the cosmonauts before launching as well as during flight and during landing; it covers the performance of the cardiovascular and the respiratory system under dynamic conditions. We have adapted for this an apparatus called 'physiotest,' developed at the Military Institute of Aviation Medicine, which can simultaneously record parameters of seven states of a human organism under dynamic conditions. Its data can be displayed on an oscillograph screen, recorded on a magnetic tape, indicated digitally on a luminous panel, and transmitted to a computer for a precise analysis of changes in the cardiovascular and the respiratory system. The apparatus can be connected to a moving track [exercise machine] and to a cyclergometer for authentic regulation of the heart activity, track speed, inclination angle, etc. We will use this for pre-launch tests of the Soviet and Polish cosmonauts, for post-landing tests, and during the debriefing period.

The subsequent two experiments we will perform jointly with Soviet specialists, as a continuation of Soviet research. The object of one of them is to test the responses of man's constitution to the use of 'Czybis' [Chibis] decompensating suits in the state of weightlessness. The other experiment deals with determining blood distribution under zero-gravity conditions. These two items are interrelated and they constitute a certain entity in an ongoing test series whose aim is to solve a problem of great importance to the future of cosmonautics; namely, the preparation of a man for long space flights.

Jointly with the Soviet Union and Czechoslovakia, we will study the heat transfer in an organism under zero-gravity conditions. The absence of air movement may inhibit heat transfer in some parts of the human body. Without dwelling on this subject, let me just point out that we have undertaken an experiment involving the use of temperature-holding sensors. Another Soviet-Polish-Czechoslovak experiment in the field of space medicine concerns oxygen processing under zero-gravity conditions. Oxygen depletion is measured here polarographically.

In the field of space psychology we have developed, jointly with the Soviet Union, a 'log book' which we hope will give us an objective assessment of the effects of various factors on man's performance during a long orbital flight. The data recorded in this diary will include fitness for activities under zero-gravity conditions, ability to communicate with other members of the crew, processes involved in visual perception, work and rest rhythms, use of medications and their effect upon the feeling of well-being, etc. In the field of psychology we will also perform tests dealing with recreation. On the basis of research on individual interests and tastes of cosmonauts, Col. Romuald Bloszczynski (Director of the Department of Psychophysiology at the Military Institute of Aviation Medicine) has prepared a four-hour recreation program and put it on video tape. This program, when followed during flight, will make it possible to evaluate the effects of weightlessness on the human body.[8]

RENDEZVOUS WITH A SPACE STATION

The Soyuz-30 spacecraft, bearing the call sign *Kavkaz*, was launched from the Baykonur Cosmodrome at 6:27 p.m. (Moscow time) on 27 June 1978. Although the West paid little attention to the first flight of a Polish cosmonaut, there was great enthusiasm throughout Poland. Celebrations by the Polish people underlined the importance that Moscow gave to the propaganda benefits of having socialist partners participate in a joint program of space research. Hermaszewski became an instant hero in his country and was compared in the press to the legendary Jan Twerdlowski, the 16th century Polish nobleman of children's literature who crowned his career as an Alpinist with a flight to the Moon on the back of a rooster.

Poland marked this historic occasion with posters, badges and medals. Television and radio stations carried news of the progress of Soyuz-30 as it headed for the space station. At 8:08 p.m. the following day the spacecraft docked at the aft port of the Salyut-6/Soyuz-29 orbital complex. The station itself had been launched on 29 September the year before, and then visited by three two-man crews between December 1977 and March 1978. Soyuz-29, crewed by Vladimir Kovalyonok and Aleksandr Ivanchenkov had linked up with it just ten days earlier, on 17 June.

Klimuk and Hermaszewski aboard the Salyut-6 space station. (Photo: Author's collection)

Once all the necessary checks had been made, the Soyuz-30 crew opened their hatch and moved into the station to be greeted by the two residents. There was the traditional greeting consisting of bread and salt, along with a little fruit juice in glasses that they could clink for good luck. After a familiarization tour of the station the new cosmonauts began that day's program of scientific experiments.

One test which all four crewmembers worked on was the Sirena experiment that used the Splav-1 smelting furnace to produce a crystal of cadmium-tellurium-mercury semi-conductor under weightless conditions. Sirena was one of the experiments which the Polish Institute of Physics had proposed for the Interkosmos program in May 1977, at the First Conference on Problems of Space Technology in Moscow. This experiment was performed in collaboration with the Institute of Space Research at the USSR Academy of Sciences in Moscow, and was noteworthy for being the first technological experiment in history to be performed with such a ternary compound in space The following tasks were carried out:

- The study of the connection between alloy uniformity and the transfer of a mass in zero gravity.
- The study of crystallographic structures.
- The magnitude of crystal growth and other observations.

The crystallization experiments would eventually yield 47 grams of cadmium-tellurium-mercury semiconductor for use by infrared detectors on board the station. At 50 percent the yield of the process was far greater than the 15 percent achieved by experiments on Earth.

Another Splav test was conducted on 3 July and materials from this experiment would be brought back to Earth when the visitors undocked and returned two days later. Kovalyonok and Ivanchenkov would continue these experiments with Sirena-2 and Sirena-3.

The two crews undertook photography of the Earth's surface and of aurora phenomena (in this case the 'northern lights'). However an indication of the relatively low priority given to these 'guest cosmonaut' missions was that the activities of the Soviet/Polish crew were often curtailed in order not to interfere with the schedule of the residents.

Hermaszewski also participated in medical experiments which measured lung capacity and the heart during exercise and in a pressure suit. Tests carried out on 1 July showed that both visiting cosmonauts were in good condition, with Hermaszewski's heart rate recorded at 55 and his blood pressure at 125/60; the corresponding figures for Klimuk were 75 and 120/70.

Television images from Salyut-6 were beamed down to the ground. The Soyuz-30 crew are joined by resident crewmember Aleksandr Ivanchenkov (left of photo, behind Hermaszewski) (Photo: Author's collection)

As with all of the international crews, both Klimuk and Hermaszewski had been trained in the use of the MKF-6M multispectral camera. This had been developed in East Germany and had six lenses for photographing the Earth's surface from space using a variety of filters. It was not simply a matter of point and shoot; first the camera had to be primed, then the Salyut station had to be oriented to aim the camera directly at the ground when the photograph was to be taken.[9] The training to operate the camera usually took place on a Tu-134 flying at an altitude of six miles. From the station Hermaszewski photographed Poland in coordination with aircraft flying the same ground track at lower level, but often bad weather limited these photo sessions.

One experiment in which all four cosmonauts participated was called Smak or Vkus (both meaning Taste) which investigated a spacefarer's ability to differentiate between certain food items, and why some which were enjoyable on Earth became far less palatable in space.

The MKF-6M camera which was installed on Salyut-6 and (below) a postage stamp issued to commemorate its use aboard the station. (Photos: Spacefacts.de and Deutschen Post der DDR)

SAFELY BACK ON EARTH

The Soyuz-30 crew's work was completed on 5 July and they loaded their spacecraft with material from the scientific experiments that needed to be returned to Earth. At 1:12 p.m., after the two cosmonauts had taken their leave of Kovalyonok and Ivanchenkov, Klimuk undocked from the station. The landing occurred 3 hours and 18 minutes later, some 190 miles south-west of Tselinograd in the north of the central Asian republic of Kazakhstan after a mission lasting 190 hours and 3 minutes. Once they had emerged from their spacecraft the two men said they felt good, although Hermaszewski reported he was feeling slightly sick in the stomach and was a little unsteady on his feet.

Their mission at an end, Hermaszewski and Klimuk stand in front of the charred spacecraft. (Photo: Author's collection)

During the Soyuz-30 mission, back-up pilot Jankowski had acted as a "consultant to the Senior Ground Flight Controller" (CapCom) in the Flight Control Center at Kaliningrad, near Moscow. After the successful completion of the flight, he was promoted one step to colonel and was awarded the Commanders Cross of the Polonia Restituta (otherwise known as the Commodore Cross of the Order of Poland's Resurgence) in recognition of "his services in the preparation of the flight and exemplary fulfillment of his tasks as consultant to the Flight Controller."[10] He later retired from all duties.

Although he mostly seems to avoid the spotlight these days and shuns media interviews, on occasion Col. Jankowski does offer an opinion on space-related endeavors. In one brief radio interview he was asked about the question of space tourism and whether, if he had the necessary funds, he would embark on such a flight.

"Absolutely not," was the unapologetic response. "And certainly not in the role of a passenger on a space taxi. I wonder what these people actually want to prove by flying into space. Victory over fear? The genius of the human being? To me this all seems to be very

A recent photo of Jankowski. (Photo: Author's collection)

strange. There are other ways to demonstrate a person's humanity. I heard about a man who became well known for building houses for the poor. This to me has far more value than just flying in space."[11]

There is a further element to add to the story of Zenon Jankowski. According to Douglas Hawthorne, writing in *Men and Women of Space*, "Unofficial philatelic information from Poland seems to indicate that Jankowski, not Hermaszewski, was chosen by the Soviet Union to be the first Pole in space. The Polish government printed two stamps honoring Jankowski as part of the 1978 Interkosmos program [bearing the words *Pierwszy Polak w kosmosie* (the first Pole in Space)]. The day before the flight, the Polish philatelic agency Ars Polona received word that Hermaszewski was the first Polish cosmonaut, and new stamps were printed. The incorrect stamps were destroyed, but some apparently escaped the shredding machines."[12]

However, this lingering rumor has since been proven to have been just a fanciful fiction. Under existing protocol, the Interkosmos prime crew was always approved at the time of the formation of the prime and back-up crews several months ahead of the flight. This would obviously change if there was a serious problem requiring the crews to change positions. An official announcement of the crew that had been selected to make the flight was then issued just two days ahead of the launch date, which in this case confirmed the prime crew pairing of Klimuk and Hermaszewski.

Once the post-flight celebrations and tours had died down, Hermaszewski returned to service in the Polish Air Force and was promoted one step to the rank of lieutenant colonel. When martial law was imposed in Poland in December 1981 he was undertaking studies in Moscow at a military academy. Without any prior notification, and to his utter surprise, he was informed that he had been assigned as one of 20 members of Poland's Military Council for National Salvation, a military quasi-government. He was ordered to return immediately, arriving in Warsaw on the evening of 13 December 1981. Hermaszewski later said that he had actually learned from television that he was a member of the military government. He had not been asked whether he wished to take part, not advised that he had been appointed, and not asked whether he consented to his appointment. The following year he graduated from the K.E. Voroshilov Military Academy in Moscow and subsequently served for three years as the Chief of the Shkola Orlyat High Aviation School. He was also selected as the President of the Polish Astronautical Association in 1983. After a position in the Defense Office of Poland, he served from 1987–1991 as a commander of the 'School of Eaglets' in Deblin, being promoted to the rank of general in 1988.

In 1991, Hermaszewski became the Deputy Commander of the Polish Air Forces and, subsequently, an inspector in the General Staff of the Polish Armed Forces.

Brig. Gen. Hermaszewski is currently retired and a member of the Association of Space Explorers and Polish Academy of Sciences. He still holds a few official Polish records recognized by the International Air Sports Federation, including Altitude of Flight, Velocity of Flight, and Endurance of Flight. He and his wife Emilia are currently living in Warsaw.

In 1999, NASA payload specialist James Pawelczyk (STS-90), the first American astronaut of full-blooded Polish descent, traveled to Poland and met the first Polish cosmonaut, Miroslaw Hermaszewski. (Photo: PAP/M. Trembecki)

His many awards include: Pilot-Cosmonaut of the Polish People's Republic; Hero of the Soviet Union; Soviet Order of Lenin; Polish Order of the Cross of Grunwald, First Class; Polish Academy of Sciences Nicolaus Copernikus Medal; Order of the Gold Cross of Merit (all awarded 1978), plus various other orders and medals.

The Soyuz-30 capsule is currently on display at the Muzeum Polskiej Techniki Wojskowej (Museum of Polish Military Technology) in Warsaw.

EXPERIMENTS CARRIED OUT ON SOYUZ-30

Sirena: This consisted of two experiments aboard Salyut-6, one in the Splav furnace and the other in the Kristall furnace. The aim of the experiment in the Splav furnace was to study the process of directed crystallization of semiconductor materials from the liquid phase in zero-g. For this purpose a cylindrically shaped crystal of the triple semiconductor mercury-cadmium-tellurium (HgCdTe) was placed in a quartz ampoule and melted in the furnace, after which it was cooled and allowed to solidify and then removed from the furnace and returned to Earth.

HgCdTe semiconductor alloys are difficult to create on Earth because the three elements have significantly different atomic weights and convection processes (resulting from gravity) make it difficult to ensure those alloys remain homogenous. HgCdTe semiconductor crystals have been studied for many years by the Physics Institute of the Polish Academy of Sciences. Their main application is in infrared detectors, more particularly in detectors that can detect radiation with a wavelength of about 10 micrometers, which is in the atmospheric 'window' where water vapour and other components do not absorb this radiation. Such materials can also be used in semiconductor lasers.

It was expected that HgCdTe crystals obtained in zero-g would be more homogeneous and have a more perfect crystalline structure. The experiment was prepared in collaboration with the Institute of Space Sciences (IKI) of the Russian Academy of Sciences in Moscow.

The experiment in the Kristall furnace was to study the crystallization of semiconductors consisting of lead-selenium-tellurium (PbSeTe) or cadmium sulphide with small additions of selenium (PbSnTe) from the gaseous phase. The materials were placed in a quartz ampoule, heated to vaporize them, and then allowed to form a monocrystal layer on a cold plate. After the experiment, the ampoule was extracted from the furnace and returned to Earth.

Triple semiconductors based on lead are used in infrared semiconductor lasers. They are only effective if they have a homogenous structure and this is difficult to produce on Earth. The experiment was prepared in collaboration with IKI.

Opros (Questionnaire): An experiment designed to study how comfortable life in space is and to evaluate work efficiency in space. For this purpose the Military Institute of Aviation Medicine in Poland teamed with Soviet specialists to develop an appropriate questionnaire. The cosmonauts were to periodically answer questions on a scale of one to five concerning their sleep patterns, the taste of food, adaptation to the conditions in space, the interactions between international crew members, and so on.

Vkus (Taste): This experiment investigated changes in taste patterns in zero-g by electric stimulation of the taste nerves employing a Polish device called an "elektrogustometr" that generated a sawtooth voltage, making it possible to produce an alternating current between electrodes (from 0 to 300 microamperes) for a digital measuring system. One electrode was placed on the tongue and another on a hand. The electric impulses would generate either a metallic or a sour taste. The system measured the current required to cause a sense of taste.

Dosug (Leisure time): The Military Institute of Aviation Medicine in Poland teamed with Polish television and Soviet specialists to record several four-hour television programmes. The cosmonauts had to judge them by making notes in a special section of their on-board journals. The objective was to more effectively organize the leisure time for future crews.

Kardiolider: The crew measured changes in the heart rate employing the Kardiodiler-D01 instrument that was developed by the Factory of X-Ray and Medical Equipment in Warsaw. It was a portable device weighing 200 grams that had an autonomous power supply. When electrodes were placed on the body, it could measure changes in the heart rate in the range from 60 to 180 beats per minute. The first part of the experiment was conducted during the second flight day, while the cosmonauts are wearing the Chibis vacuum suit. If there were any indications that a cosmonaut's condition was deteriorating and his heart rate exceeded 100 beats per minute, the experiment would be automatically stopped.

The second part of the experiment was conducted during the third day of the flight while the cosmonauts were doing physical exercises on the VEL-1 stationary bicycle (known as a cyclergometer). In the first training mode their heart rates reached 130 beats per minute and in the second mode they reached 150 beats per minute. If the heart rate exceeded the given parameters in the first mode, the cosmonaut would slow down the speed. If the heart rate exceeded the given parameters in the second mode, the cosmonaut would stop the exercise.

The experiment was designed to help identify efficient ways to prevent cardiovascular deterioration during long-duration missions and to help maintain work capacity during all stages of the flight.

Other experiments: The visitors continued the 'Kislorod', 'Teploobmen' and 'Khlorella' experiments begun by the Soviet-Czech crew, took photographs of the Earth in the interests of the national economy, and took part in a medical experiment to study the reactions of the cardiovascular system to simulated gravity in the Chibis suit.

The book *Orbity Sotrudnichestva* also mentions a medical experiment that investigated blood circulation using devices known as Polinom-2M, Reograf, and Beta. These had been introduced on Salyut 4. Reograph was an apparatus to monitor the distribution of blood by measuring blood flow in the head, torso and extremities. Polinom-2M was a unit to monitor the performance of the heart. In addition to general cardiograms, it could isolate the major blood vessels and monitor the phases of the cardiographic cycle. Beta was a multifunction medical unit that complemented the electrocardiogram functions of Polinom-2M, but it also measured the lung capacity and recorded seismograms to determine the rhythm of the heart and the force of blood pumping.

REFERENCES

1. Polskie Radio DLA Zagranicy interview/article, "Astronaut recalls 'miraculous' survival of WWII massacre," dated 4 July 2013. Available online at: *http://www.thenews. pl/1/10/Artykul/140336,Astronaut-recalls-miraculous-survival-of-WWII-massacre*

2. *Voice of Sevastpol* article, "Europeans, Seems You Are Shameless," dated 4 August 2014, online at: *http://en.voicesevas.ru/news/yugo-vostok/2578-europeans-seems-you-are-shameless.html*

3. Polskie Radio DLA Zagranicy interview/article, "Astronaut recalls 'miraculous' survival of WWII massacre," dated 4 July 2013. Available online at: *http://www.thenews. pl/1/10/Artykul/140336,Astronaut-recalls-miraculous-survival-of-WWII-massacre*

4. Russian cosmonaut list (Poland), available online at: *www.astronaut.ru/as_polsk/text/ quelle.htm*

5. Paper, *Translations on Eastern Europe, Scientific Affairs, No. 597*, JPRS 71699, 17 August 1978

6. Gordon Hooper, *The Soviet Cosmonaut Team, Vol. 1, Background Sections,* GRH Publications, Suffolk, UK, 1990

7. Radoslaw Nawrot, online *Gazeta* article "Poznaniak, ktory nie polecial w Kosmos" (Poznian who did not go into space). Online at: *http://wiadomosci.gazeta.pl/ kraj/1,34309,5405329.html*

8. Article by Col. Stanislaw Baranski, "Translations on Eastern Europe Scientific Affairs, No. 597," JPRS, 17 August 1978

9. David Harland, *The Story of Space Station Mir*, Springer-Praxis Books, Chichester, UK, 2005

10. Gordon Hooper, *The Soviet Cosmonaut Team, Volume 2: Cosmonaut Biographies*, GRH Publications, Suffolk, UK, 1990

11. Radoslaw Nawrot, online *Gazeta* article "Poznaniak, ktory nie polecial w Kosmos" (Poznian who did not go into space). Online at: *http://wiadomosci.gazeta.pl/ kraj/1,34309,5405329.html*

12. Douglas B. Hawthorne, *Men and Women of Space*, Univelt Inc. Publishers, San Diego, CA, 1992

4

Sigmund Jähn and Soyuz-31

The son of Paul Jähn, a sawmill worker, and his wife Dora, a seamstress who also worked part-time for the forestry association, Sigmund Werner Paul (Sig) Jähn was born on 13 February 1937 in the village of Rautenkranz, in the Vogtland area of what would become East Germany – also known as the German Democratic Republic (GDR). He attended elementary school in the village from 1943 to 1951.

As a youth, Jähn did not have a clear idea of what he wanted to do, with vague ideas of becoming an engine driver or a forester. At the age of 14 he left school and completed an apprenticeship as a book printer and typographer at the branch of the Falkenstein national book printing works in the nearby town of Klingenthal, where he gained a skilled workers' certificate.

After briefly working as a pioneer director at a school in Hammerbrucke-Vogtland, in 1955 he volunteered for military school, the forerunner to the Air Forces of the National People's Army, and on 26 April began training as a National People's Army cadet pilot at the Franz Mehring Officer School in Kamenz, where he was later described as a model cadet. He became a member of the Free German Youth, and in 1956 joined the Socialist Unity Party.

Jähn next undertook officer training, first in Kamenz and then from May 1957 in nearby Bautzen. After finishing two years of basic training he graduated in 1958 and received his commission as a second lieutenant. He then became a pilot with a fighter squadron in the NVA (*Nationale Volksarmee*). By then he had met his future wife Erika. She was training as a fitter and later qualified as a technical draftsman. After a brief courtship they were married. Their first child, Marina, was born in 1958. From 1961 to 1963 Jähn was deputy commander for political and ideological instruction within his unit, and in 1965, at the age of 26, he was assigned as his squadron's head of air tactics and gunnery.

In 1966 Jähn, now the father of a new-born second child, a daughter they named Grit, was sent to the Soviet Union for additional training at the Yuriy A. Gagarin Military Academy in Monino, near Moscow, where he proved to be an exemplary student by receiving 13 ratings of "excellent" and 8 of "good" in his 21 training subjects.

In an interview for *Spiegel Online*'s Olaf Stampf in 2015 to mark the 50th anniversary of the orbital mission of Yuriy Gagarin, Jähn was asked if he had ever met the history-making cosmonaut.

© Springer International Publishing Switzerland 2016
C. Burgess, B. Vis, *Interkosmos*, Springer Praxis Books, DOI 10.1007/978-3-319-24163-0_4

People's Army cadet Sigmund Jähn. (Photo: German Spaceflight Exhibition Morgenröthe-Rautenkranz)

Young married couple Erika and Sigmund Jähn. (Photo: German Spaceflight Exhibition Morgenröthe-Rautenkranz)

"No, I was too young. But when he died tragically seven years after his space flight, I was studying nearby, at the Soviet Air Force's military academy. I attended the funeral service in Moscow, and I saw for myself how people waited in line for hours, with tears in their eyes, to pay their respects. It was genuine grief, not propaganda."

He was asked to comment on various theories about Gagarin's plane crash in 1968. Was it just an accident? Or suicide? Even murder?

"Those are nothing but rumors. Anyone who has never flown a MiG-15 UTI would be better off keeping quiet. As a flight instructor, I've flown this model often and I've studied similar crashes. Even at high speeds, it is possible for the fighter jet to stall, especially in clouds. Also, Gagarin had hardly flown since his space flight. I agree with the conclusion that the plane was at an unstable inclination, and there wasn't enough time to pull out of the dive."

When asked if Gagarin's pioneering flight had inspired dreams of traveling into space himself, Jähn responded, "No, that was all very far away. Of course we, as fighter pilots, were glad that people like us were needed for space travel. But that was all."[1]

MiG-17 fighter pilot in the GDR Air Force. (Photo: Mitteldeutscher Rundfunk)

After graduating from the Military Academy in 1970, Jähn returned to East Germany. He then served for five years as Inspector for Flight Safety at the Air Defense Command of the NVA in Strausberg, investigating incidents and accidents involving military aircraft. For his exemplary service to the air force he was awarded the title of Distinguished Military Pilot of the German Democratic Republic. By this time he had accrued more than 1,000 hours in the air. It was at this stage of his aviation career that he was offered the opportunity of a lifetime – the chance to fly into space.

A GOLDEN OPPORTUNITY

Following discussions with Eastern bloc representatives in Moscow in July and September 1976, it was agreed that citizens of the socialist countries participating in the Interkosmos program would assist in selecting suitable candidates to accompany Soviet cosmonauts on flights in Soyuz spacecraft to a Salyut space station. At the 14 September meeting, it was decided the missions would occur in the period 1978–1983. It was also decided to conduct the training in the Yuriy Gagarin Cosmonaut Training Center in Star City, near Moscow, starting on 1 December 1976, but that snippet of information wasn't released to the general media.

As there was insufficient time to perform a lengthy search for suitable candidates, it was decided in the German Democratic Republic to cut straight to surveying the annual reports which were prepared on pilots of the GDR Air Force and then arrange preliminary medical examinations for those who made the cut.

After the personnel files had been screened, the next stage in the selection process was a personal interview conducted by Lt. Gen. Wolfgang Reinhold, commander of the GDR Air Force and the Deputy Defense Minister of the GDR. Several hundred prospective candidates from a number of units across the country were subjected to this interview phase. When Jähn received his summons he had no idea what it entailed.

"It was all secret, really top secret. One morning I was ordered to report to the head of the air force. I suspected nothing, just thought: Did I do something wrong? There were about a dozen other experienced pilots sitting next to me. We had all done our training at the Soviet military academy."

As each man was summoned to Reinhold's office, he was asked if he was interested in participating in a joint international space flight. They were given two days to think it over, but as Jähn later stated, he didn't even need two minutes. "Finally, they let the cat out of the bag. I said yes immediately."[2] By the conclusion of this phase of the process, 30 qualified candidates had been identified.

At the end of September 1976 the remaining candidates were sent to the Central Aviation hospital, where doctors conducted an extensive medical examination on each man. Particular attention was paid to the cardiovascular system, lung capacity, and reflexes. The men were then questioned on their knowledge of physics, mathematics, and the Russian language.

According to the authors of the Polish website *Loty Kosmiczne*, by the conclusion of this phase of the operation on 1 October 1976 there were nine candidates remaining,[3] namely:

- Rolf Berger
- Heinz Boback
- Eberhard Golbs
- Christian Haufe
- Sigmund Jähn
- Walther Jehnichen
- Eberhard Köllner
- Peter Misch
- Wolfgang Wehner.

According to Eberhard Golbs in a recent interview, Walther Jehnichen, Peter Misch, and Wolfgang Wehner "were only present on certain days or up to one or two weeks, and did not belong, for the most varied reasons, to the pre-selection group of the GDR". Golbs also had two vivid recollections from that time. "First of all, I was the oldest amongst the candidates (41 years of age), which was considerably above the age limit specified by the Soviet side. Secondly, it was clear to everyone that only one candidate would fly. This did not lead to [the competitive] 'cut and thrust' that one might have expected; in fact there was a spirit of good camaraderie in which everyone invested his assets to ensure that 'one of us is going to make it'."[4]

The next phase of the process involved sending a delegation of medical specialists from Moscow, accompanied by cosmonaut/physician Col. Dr. Vasiliy Lazarev. They conducted their own intense physical and psychological tests and finally came up with four names,[5] all with the GDR-AF rank of lieutenant colonel:

- Rolf Berger
- Eberhard Golbs
- Sigmund Jähn
- Eberhard Köllner.

A group photo of the Czech, GDR, and Polish candidates with accompanying physicians in front of a bust of Yuriy Gagarin in Star City. From left: Col. Dr. Hans Haase (physician), Eberhard Köllner, Eberhard Golbs, Henryk Halka, Miroslaw Hermaszewski, Zenon Jankowski, Tadeusz Kuziora, Krzusztof Klukowski (physician), Ladislav Klima, Rolf Berger, Vladimir Remek, Sigmund Jähn, Antonin Dvorak, (unknown Soviet supervisor), Oldrich Pelcak and Michael Vondrousek. (Photo courtesy of Eberhard Golbs)

The four GDR candidates: Eberhard Köllner, Eberhard Golbs, Dr. Haase, Rolf Berger and Sigmund Jähn. (Photo courtesy Eberhard Golbs)

On 10 November 1976 the four candidates were sent to the Gagarin Cosmonaut Training Center, where they were presented to members of the Chief Medical Commission for a final evaluation. Each man also had to endure an open session before the National Commission, specifically created for the selection of cosmonaut candidates. In this, a man's professional skills and good health were taken into account, as well as his ideological convictions, moral stability, intellectual breadth, and other qualities that were deemed to be typical of a man of socialist society.

After a stringent and exhausting series of tests, interviews, and medical examinations, on 25 November it was announced that Rolf Berger and Eberhard Golbs had been eliminated, and that Sigmund Jähn and Eberhard Köllner would begin training for the Soviet-GDR joint flight.

"The Russians had the final say," Jähn recalled, referring to the fact that a Soviet medical commission had determined the final two places, "since it was their spaceship that we would be flying. And I performed the best in the medical tests, for example when being spun in a centrifuge."[6]

TRAINING IN STAR CITY

The first group of trainees for the Interkosmos flights arrived at the TsPK (*Tsentr Podgotovki Kosmonavtov*) training center in Star City in December 1976. This group consisted of two trainees each from the first three participating nations – Czechoslovakia, Poland and the GDR. Each man arrived with his family members and they were temporarily housed in the Hotel Kosmonavt until they could be allocated more permanent accommodation in a new apartment block. The next day the Interkosmos candidates reported for their first training session. These sessions were divided into four periods. Overall the training would last for approximately 18 months.

The 2005 Springer-Praxis book *Russia's Cosmonauts: Inside the Yuri Gagarin Training Center*, by Rex Hall, David J. Shayler and Bert Vis offers an excellent insight into the four sets of training sessions, here reproduced:

First Training Period (6–7 months)

Most candidates found this to be the hardest part of the training program, with lectures in basic theoretical education covering astronomy, navigation, mathematics and the Russian language. During these months, the candidates visited several factories responsible for the production of space flight hardware and systems and, occasionally, sites of national cosmonautics importance, such as the house of Konstantin Tsiolkovskiy and the museum of space achievements. They also took a parachute jumping course and participated in fit tests for Sokol pressure garments. Their training advisors included former cosmonauts Leonov, Volynov, Yeliseyev and Lazarev. Towards the end of the first training period a special training group of veteran cosmonauts was formed to prepare in parallel for the role of Soviet commander on the joint flights.

Second Training Period (over 2 months)

For the Interkosmos candidates, this second period consisted of lessons on the basics of spacecraft technology, and it proved difficult even to the ardent students of technology. Used to the 'aeronautics' of an aircraft, they found the 'astronautics' of a spacecraft a completely different methodology of flying. Meanwhile, the experienced cadre of Soviet commanders worked on their role of taking a Soyuz to and from a Salyut station almost single-handed. Towards the end of this period, the Interkosmos candidates were 'paired' with their Soviet commanders and identified as the primary or back-up candidate. The chief of the training center, former cosmonaut Lt. Gen. Georgiy Beregovoy, announced the crew assignments based on training performance and in sequence to mark significant events in the respective communists country's history, as well as for requirements in the Salyut manifest. However, poor performances, illness, or better performance from the assigned back-up crew could result in a change of assignments, so all candidates had to be sure to put in 100 percent effort if they wanted to keep their flight.

Third Training Period (about 4 months)

The training program became even busier during the third period. Having been teamed with an experienced Soviet commander, this was when the Interkosmos candidates put theory into practice while learning to operate as a crew with numerous sessions of up to six hours per day in both the Soyuz and Salyut simulators, coupled with endurance tests in the decompression chamber and centrifuge. To achieve this meant long working days, extending an hour or two after the 'official' TsPK working day finished at 2100 hours. The paired crew also socialized together, thus getting to know each other's families and accompanying each other on off-duty social events and visits. Though becoming closer as a crew, the Interkosmos cosmonauts were still restricted in what they could or could not handle inside the spacecraft.

Fourth Training Period (about 4–6 months)

The final training period included specific mission training and familiarization with the scientific instruments and experiments supplied by the relevant Interkosmos nation. This period covered repeated simulations of key mission events, such as launch, rendezvous, docking, work aboard the space station, undocking, and landing. In addition, the crews completed water landing and water egress survival training in the Black Sea, and other wilderness survival training in the event of emergency landing situations. The physical endurance tests were stepped up, and simulator runs became daily as the scheduled date of launch approached. About one month prior to the planned mission, final exams were taken to confirm the primary and back-up positions.[7]

Two weeks prior to the Soyuz-31 launch, by the normal routine, the four crewmembers flew down to the Baykonur Cosmodrome located in the desert steppe of Kazakhstan. Once there, they had familiarization sessions with their actual spacecraft and kept informed of the sequence of events associated with the preparations for launch.

A few days prior to launch day, again following a much-practiced routine, the crews were confirmed and authorized fit for flight. It was at this time, much to their relief, that Valeriy Bykovskiy and Sigmund Jähn were confirmed as the prime crew for the Soyuz-31 mission.

The prime and back-up crew training together. (Photo: Author's collection)

The roll-out of the rocket that will carry Soyuz-31 into space. (Photo: Author's collection)

AHEAD OF LIFT-OFF

At 5:51 p.m. Moscow time on 26 August 1978, the Soyuz-31 spacecraft lifted off from launch pad 1/5 at Baykonur atop a Soyuz-U rocket. With a call sign of *Yastreb* (Hawk), Valeriy Bykovskiy and Sigmund Jähn were headed for a rendezvous and docking with the Salyut-6 space station and its then-resident crew of Vladimir Kovalyonok and Aleksandr Ivanchenkov, who had been aboard since Soyuz-29's arrival on 15 June.

Just a few days before he was due to be launched, 41-year-old Sigmund Jähn had become a grandfather, when his daughter Marina had a baby. However the birth was not mentioned in the media.

Prior to Bykovskiy and Jähn boarding their Soyuz, on the steps to the elevator at the bottom of the rocket, the East German press had a chance to ask Jähn about his upcoming mission, presumably with questions they had been instructed to put to him.

"How do you feel; what are your thoughts at this time?" one reporter asked.

"I want to thank everyone, who, through their work, made it possible for me to experience this," Jähn responded. "In particular the workers, scientists and engineers who were immediately involved in the training. I'm very proud that I may experience this launch as a representative of all our citizens."

From another reporter: "Comrade Jähn, in spite of the obvious excitement, you both look so relaxed. Where does this confidence come from?"

The answer was well rehearsed. "We've been prepared well, we trust the technology, and we are sure that we will fulfill the program successfully."

Bykovskiy was not a willing partner to all of this political carry-on, and he made a less serious comment in which he said that at this time it was certain he wouldn't be seeing a cigarette for a week.

Later, when strapped into their Soyuz spacecraft awaiting the launch, Jähn gave another statement that was shown on television, saying: "I dedicate my flight to the 30th anniversary of the founding of the German Democratic Republic, my socialist fatherland." Jähn had been ordered to say that by his superiors in Berlin, and in fact a colonel had earlier provided him with a piece of paper specifying the exact text that he was to use. Nothing, it seemed, was allowed to be spontaneous.

Elsewhere on the morning of the launch, the chief editors of all newspapers and radio stations in the German Democratic Republic received three sealed envelopes marked with numbers which had been carefully prepared and sent by the Central Committee of the SED (Socialist Unity Party). These could only be opened after receiving telephone instructions to do so. The first envelope contained photographs, text, and a headline: "The First German in Space – a Citizen of the GDR." Following the successful launch the telephone call was made and the contents were immediately read out and published. The other two envelopes were later recovered by messengers from the Central Committee and destroyed. They contained carefully worded statements that were to have been published in the event of a fatal incident at launch or if technical difficulties caused the spacecraft to come down on so-called enemy territory.

The Soyuz-U rocket stands ready for lift-off, and the moment of ignition. (Photos: Author's collection)

RENDEZVOUS, DOCKING AND HARD WORK

Following a flawless launch Soyuz-31 slipped effortlessly into orbit. Jähn later described the early phase of the ascent as "bumpy", and recalled that veteran cosmonaut Aleksey Leonov once described a Soyuz launch to him as "driving over a cobblestone road in a car that has square wheels".

The day after lift-off, at 7:38 p.m. Moscow time, Bykovskiy completed a seemingly fault-free docking at the aft port of Salyut-6. Soviet television gave a view of the link-up. When it came time to open the Soyuz hatch, however, it wouldn't budge. Each of them tried without success. For a while it looked as if the flight might end in failure and an ignominious return to Earth, but the men joined forces, took a firm stand against the wall of the orbital module, and finally met with success as the hatch reluctantly opened up.

The Salyut-6 space station as photographed by the crew of Soyuz-31. (Photo: Spacefacts.de)

Jähn was the first to enter the space station. He floated into the main compartment and shook hands with Kovalyonok and Ivanchenkov, then kissed both men on the cheek in a traditional greeting. The visitors had come with mail and gifts, the latter including recent copies of the newspapers *Pravda* and *Izvestia*, and an inscribed watch for each man. Jähn also presented them with a book written in Russian about the German Democratic Republic. Soviet television showed the four men smiling and laughing. Announcing, "Now we invite you to our table," Kovalyonok offered them the traditional welcome meal of bread and salt.

All too soon it was time to get down to work, because the new arrivals had a demanding seven-day program of scientific research scheduled.

One of the main assignments for the international crew was the Biosphere experiment. This was a study of the Earth's surface using the multi-spectral MKF-6 M camera and the Pentacon-6 M and Praktica EE2 hand-operated cameras supplied by the Pentacon factory in Dresden, East Germany. Apart from photography, they would make visual observations of air pollution, study meteorological processes, including the formation and movements

The four cosmonauts aboard Salyut-6. From left: Bykovskiy, Ivanchenkov, Kovalyonok and Jähn. (Photo: Author's collection)

of clouds, and ocean currents. Principal areas of observation would include the Tuzgulu salt lake in Turkey, the Kock-i-Sultan volcano in Pakistan, the Bahama and Canary Islands, Cape Khafun in Somalia, Mauritania, the Alps, and the Amazon River delta. They were also to conduct micro-organic experiments and investigate the properties of organic polymers. And of course, the four crewmen would be using electrocardiograms and other equipment to test the response of their cardiovascular system to the space environment.

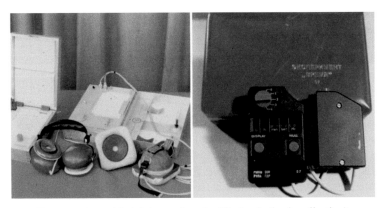

Instruments used in joint experiments. (Photo: Author's collection)

Some of the experiments would be carried out each day and take just a few minutes, but others, such as the one using the Splav and Kristall installations to obtain metallic and non-metallic materials in weightlessness would be performed only once but take several hours. Scientists from the GDR Academy of Sciences and Berlin's Humboldt University had drawn up a program for producing semi-conducting materials and different glasses in space in this series of experiments.

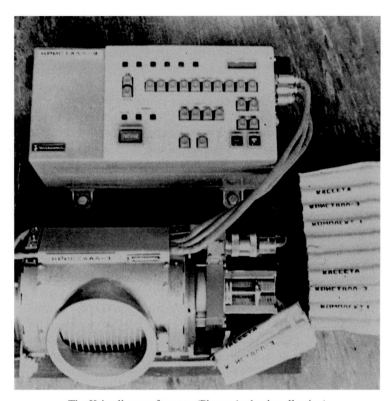

The Kristall space furnace. (Photo: Author's collection)

In 1998 Jähn shows off the Sandmännchen puppet he carried into space on his flight. (Photo: Bundesarchiv)

There was one touch of serendipity during Jähn's tenure aboard Salyut-6. He had taken with him a small figure of the Sandmännchen, a popular children's puppet character from television that had been inspired by the 'Little Sandman', a character from a story by Hans Christian Andersen. He was asked if he took the little puppet on his flight as some sort of talisman.

"That wasn't my own personal talisman, it was actually a very official assignment," he replied. "I was supposed to shoot footage for a children's program while in orbit. To that end, the Sandmännchen even wore his own spacesuit, specially made. But things didn't go according to plan. The commander, Vladimir Kovalyonok, also had a doll along, the Russian mascot Masha. We had fun pretending to marry the two dolls, but our silliness backfired. East German television couldn't really present children with a married Sandmännchen. So the footage was never broadcast."[8]

When asked post-flight what impressed him most about being in space, Jähn responded: "Looking at the Earth, the Northern Lights, the fragile-looking atmosphere, the sunrises that followed quickly one after another – these images are etched into my memory forever. From space, one thing is clear – this planet isn't so big that humans could not destroy it with their greed for profit."[9]

Following a week of intense experimentation and studies, Bykovskiy and Jähn were set to return to Earth. They would not use the Soyuz-31 craft for the journey home, but Soyuz-29. This would then leave the residents with a 'fresher' spacecraft. This change-over meant that they would have to swap items from one spacecraft to the other, including their Sokol suits, contoured seat couches and personal effects. Once this was done, Bykovskiy and Jähn filled Soyuz-29 with their scientific results and other items. The two craft would then have to be carefully trimmed and balanced to account for the different weight of the occupants and the items they would be taking back with them.

Just as they were packing for their return to Earth, a fly which had been the subject of an experiment escaped from its container and floated around, vainly flapping its wings in the weightless environment. Catching the creature caused the packing to take three revolutions of the globe longer than initially anticipated.

On 3 September, their work at an end, Bykovskiy and Jähn said goodbye to the residents of the station, entered Soyuz-29, then closed and locked the transfer hatches. Soon after, at 11:20 a.m., they undocked, and slowly withdrew from Salyut-6. After the hatch between the descent and orbital modules had been closed and sealed for re-entry, Jähn suddenly realized that he had left his onboard shoes behind in the orbital module. This meant he might have to walk around in his socks after landing, which would be a little embarrassing and potentially uncomfortable.

Just three hours later, their descent module was a fireball tearing through the atmosphere. Although the flight was meant to end with a relatively soft touchdown, their elation quickly turned to alarm as things went wrong when they approached the ground.

Jähn candidly admits that the landing was rough at the time, and has caused him problems ever since. "Just before impact, the commander was supposed to push the button to detach the parachute, but he slipped. Then it was too late. A strong wind caught the parachute and we were dragged across the steppe and rolled over several times. I took the worst hits, and I've had back problems ever since."[10]

The Soyuz-29 descent module lands in Kazakhstan. (Photos: Author's collection)

The capsule landed at 2:20 p.m. Moscow time, some 88 miles south-west of Dzhezkazgan in the Karaganda region of Kazakhstan after a journey that lasted 188 hours and 49 minutes.

The issue of Jähn's missing shoes was solved when a member of the recovery team lent him his training shoes.

Although the men seemed to be in good condition and were reported to "feel excellent" afterwards, the truth about their hazardous landing did not become public for some years.

The two cosmonauts relax before being flown out of the landing area. (Photo: Author's collection)

BACK HOME A HERO

Back on Earth, Sigmund Jähn became a 'star' in the GDR. He was showered with honors, and numerous schools and roads were named after him. The jubilant celebrations seemed almost limitless. The cities of Dzhezkazgan, Berlin, Karl-Marxstadt (now Chemnitz) and Strausberg all made him an honorary citizen.

Public adulation erupted when Jähn returned home with Valeriy Bykovskiy by his side. (Photo: Peter Koard/Bundesarchiv)

Inevitably, Jähn became (and was heavily promoted as) a national hero. So many images were posted of him in every town and village that people began to make up jokes about this situation. It was said that the answer to the question, "How far is it to …" was not measured in miles but so many Jähn. And then there was the Eiche-Bürgerstube, the oldest bar in the town of Pankow which posted a large picture of Jähn on the wall with some good advice for the nation's youth: "Drink within reason and you too can make it into space."

An amusing article in a *New Scientist* magazine in 1978 reported that, "It is impossible to walk anywhere in East Germany without bumping into pictures of folk hero Sigmund Jähn. One recent visitor to the GDR was taken aside by a native, who, in a conspiratorial fashion, whispered a joke about Jähn. It seems that someone turned up for work in a crumpled shirt. When called to account for his slovenly appearance he explained, 'It's Sigmund Jähn! I put on the television last night – Sigmund Jähn was on! I put on the radio – I heard Sigmund Jähn! This morning I switched on the radio – *still* Sigmund Jähn! I couldn't bear to turn on the iron – just in case!'"[11]

Among the many real honors bestowed upon Jähn was the naming of an East German cargo ship for him. The 10,520-ton ship was given the name *Fliegerkosmonaut der DDR Sigmund Jähn* until September 1990, when it was renamed simply *Sigmund Jähn*. It was eventually broken up in June 2009. In the front of Archenhold Observatory in the GDR capital of Berlin, a bust in his likeness was erected in the Grove of the Cosmonauts. The flight of the first German in space was told in a 1979 documentary film *Himmelsstürmer* produced by the state-owned *Deutsche Film-Aktiengesellschaft* (DEFA). In his home village, a German Aerospace Exhibition has been established. At its entrance stands the MiG-21 flown by Sigmund Jähn. His many awards and medals include Hero of the GDR, Hero of the Soviet Union, Order of Karl Marx, Order of Lenin, the Space Exploration Medal, and Meritorious Military Pilot of the German Democratic Republic.

Even though the Soviets had stressed that follow-up flights would only come about on a for-cash basis, Jähn was appointed chief of a department charged with selecting new East German cosmonauts. The presumption that these would come from the air force didn't meet

The German cargo ship named in Jähn's honor. (Photo: Author's collection)

with the approval of the Academy of Sciences. The forming of an East German cosmonaut contingent was a serious option, and a second flight was being negotiated when the Berlin Wall came down in November 1989 and it all ended there. It has been said that the slot the GDR was looking for in making a second flight was taken over by the West Germans, who used it to place Klaus-Dietrich Flade aboard Soyuz TM-14 for the Mir 92 mission to Mir in March 1992.[12]

From 1990, after the reunification of Germany, Jähn worked as a freelance consultant for the German Aerospace Center (DLR), formerly the West German space agency. In 1993 he also became a consultant to the European Space Agency (ESA) to assist them to prepare for the Euromir missions. He finally retired from this job on 1 March 2002.

Today, Jähn lives in Strausberg. He is still married to his beloved Erika, enjoys reading and hunting, and remains philosophical about his one and only flight into space and the fame that it brought him.

"I never wanted to be made into a people's hero," he once offered to a reporter. "I found it embarrassing. It wasn't my thing to deliver a lot of speeches. For me, being in the spotlight was more exhausting than the space flight itself. But I don't want to complain. I still get fan mail every day, requests for autographs and invitations to give talks. My wife always scolds me, because I answer almost every letter."

Asked what he sees ahead in space exploration, Jähn has some definite views. "We will certainly fly to Mars at some point. It's a suicide mission – but sooner or later humans do everything that they are physically capable of doing. I don't believe the visions of humans emigrating into space. Evolution has tied us to the Earth. I would rather walk in the woods of the Vogtland region [his birthplace] than float through the narrow tubes of a spaceship forever."[13]

In regard to the other three GDR finalists who were eliminated from selection in Moscow, Rolf Berger, Eberhard Golbs, and Eberhard Köllner, their careers enjoyed mixed fortunes.

Rolf Berger continued his studies at the General Staff Academy of the USSR from 1980 to 1982. Upon his return to the GDR he became commander of the 3rd Air Defense Division in Trollenhagen, and on 1 March 1984 was promoted to the rank of major

The two unsuccessful GDR cosmonaut candidates, Rolf Berger (left) and Eberhard Golbs. (Photos: Spacefacts.de)

general. In 1986 he was made deputy commander of the LSK/LV (*Kommando Luftstreitkräfte/Luftverteidigung,* the Air Force of East Germany). On 7 October 1988 he was promoted to lieutenant general and functioned as head of the air force until the German Democratic Republic was dissolved and Germany was reunified on 3 October 1990, at which time Lt. Gen. Berger transferred to the Bundeswehr (Federal Defense) as a civilian advisor to the air force. He died in 2009 and is buried in Strausberg, Germany.[14]

Eberhard Golbs continued as chief pilot with the German Air Force/National Defense until he was forced to retire from active duty, having reached the mandatory age limit. He became a member of the municipal council in the city of Löbau (Saxony) for the *Die Linke* (The Left) party, and simultaneously a candidate for the local council of the village of Ebersdorf. "I am committed to the interests of the districts and the repatriates and other migrants," he recently wrote for *Die Linke*. "Likewise, I take care of the development of transport in the urban area and the surrounding area." He continues to give talks on the cosmonaut selection process and his own experiences in that exercise.[15]

Despite the missed opportunity to fly into space, Eberhard Köllner said that he was not disappointed. On the contrary: "I'm completely satisfied with my life." On returning to the GDR after the Soyuz-31 mission he later became the Director of the Air Force Academy of the German Democratic Republic with the rank of colonel. There was a possibility in 1989 that he might still make a flight into space. That year discussions were taking place between the Soviet Union and East Germany to fly a second GDR mission, but this time on a month-long trip to the Mir space station. Both Jähn and Köllner were the likely candidates for the flight, with Köllner favored to take the flight. However the demise of the Soviet Union and the subsequent reunification of Germany put an end to these plans. With the reunification, Köllner refused to be transferred to the (West) German *Bundeswehr*. "Donning the uniform of the army would have meant to defect to the enemy," he said, explaining his decision. As an air force man, that – to him – would have contradicted his "basic political opinion".

A 2003 photo of Sigmund Jähn and Eberhard Köllner. (Photo: Quelle/G.Kowalski)

Köllner was initially unemployed after leaving the air force, but undertook retraining in business administration and found employment in private industries until his retirement in 2002. He remains good friends today with Sigmund Jähn, and is still recognized for his role in his nation's space adventure. He jokingly says "every now and then comes a pensioners' club who wants to hear again what I did."[16]

As to the two spacecraft Bykovskiy and Jähn occupied on their space mission, there is no information that the authors could find on the current whereabouts of the Soyuz-31 capsule in which the crew was launched. The Soyuz-29 descent module in which they returned to Earth is presently on display in the Deutsches Museum in Munich.

REFERENCES

1. Olaf Stampf, *Spiegel Online* article, "East German Cosmonaut Sigmund Jähn: Capitalism Now Reigns in Space," 12 April 2011. Online at: *http://www.spiegel.de/international/zeitgeist/east-german-cosmonaut-sigmund-jaehn-capitalism-now-reigns-in-space-a-756497.html*
2. *Ibid*
3. Loty Kosmiczne, attributed at website: *http://www.astronaut.ru/as_germn/gdr.htm*
4. Reply to author from Eberhard Golbs, 15 May 2015, received and translated by Jürgen Esders,
5. Loty Kosmiczne, attributed at website: *http://www.astronaut.ru/as_germn/gdr.htm*
6. Olaf Stampf, *Spiegel Online* article, "East German Cosmonaut Sigmund Jähn: Capitalism Now Reigns in Space," 12 April 2011. Online at: *http://www.spiegel.de/international/zeitgeist/east-german-cosmonaut-sigmund-jaehn-capitalism-now-reigns-in-space-a-756497.html*
7. Rex D. Hall, David J. Shayler and Bert Vis, *Russia's Cosmonauts: Inside the Yuri Gagarin Training Center*, Springer-Praxis Publication, Chichester, U.K. and New York, 2005
8. Olaf Stampf, *Spiegel Online* article, "East German Cosmonaut Sigmund Jähn: Capitalism Now Reigns in Space," 12 April 2011. Online at: *http://www.spiegel.de/international/zeitgeist/east-german-cosmonaut-sigmund-jaehn-capitalism-now-reigns-in-space-a-756497.html*
9. *Ibid*
10. *Ibid*
11. *New Scientist* magazine, unaccredited article, "Space Hero," issue 9 November 1978, pg. 465
12. Horst Hoffmann, *Sigmund Jähn, Rückblick ins All*, Das Neue Berlin publishers, Berlin, 1999
13. Olaf Stampf, *Spiegel Online* article, "East German Cosmonaut Sigmund Jähn: Capitalism Now Reigns in Space," 12 April 2011. Online at: *http://www.spiegel.de/international/zeitgeist/east-german-cosmonaut-sigmund-jaehn-capitalism-now-reigns-in-space-a-756497.html*
14. Wikipedia entry, "Rolf Berger (General)." Online at: *http://de.wikipedia.org/wiki/Rolf_Berger_(General)*
15. *Die Linke Stadtratsfraktion Löbau*, online at: *http://portal.dielinke-in-sachsen.de/stadtratsfraktionloebau/1288/251/*
16. Christian Siepmann, *Spiegel Online* article, "German in Space: The Eternal Second," 24 August 2008. Online at: *http://www.spiegel.de/einestages/deutsche-im-weltall-a-947812.html*

5

Bulgaria in space

In a somewhat obscure way, the history of Bulgarian cosmonautics actually dates back to 15 August 1964, and a reception held that day at the residence of the Soviet Defense Minister, Marshal Rodion Malinovskiy, then the Military Attaché at the Embassy of Bulgaria. During the reception Lt. Gen. Zahari Zahariev, commander-in-chief of the Air Force wing of the Bulgarian Army, raised with Malinovskiy his concept of a possible flight by a Bulgarian citizen aboard a Soviet spacecraft. He even suggested that possible candidates for the flight could be four brothers, all pilots, named Stamenkov, who happened to be at the event. The brothers were introduced to Malinovskiy and also to Soviet Gen. Nikolay Kamanin, head of cosmonaut training.

Malinovskiy did not treat the request at all seriously, although an article about the reception was published the following day in the *Red Star* newspaper, along with a photo of the four brothers. Bulgarian sources later identified the men as Krum, Stamenka, Karamfil and Eugene Stamenkov. However it would remain nothing more than a wistful and highly unlikely dream because back in 1964 the Soviet Union did not have a spacecraft suitable for such flights, nor the inclination to pursue such a fanciful request. So nothing ever came of Zahariev's proposal.

A DREAM OF FLIGHT

Georgi Ivanov (nicknamed Gosho) was actually born Georgi Kakalov on 2 July 1940 in the village of Mikre, located in the Varosha neighborhood of the ancient, picturesque northern Bulgarian town of Lovech, situated astride the river Osam. The circumstances of his name change would relate to his selection later in life as a cosmonaut researcher. His parents were Anastasia Georgieva, a housewife, and Ivan Kakalov, a master craftsman and a self-taught electrical technician employed by the Power Supply Administration of Bulgaria.

© Springer International Publishing Switzerland 2016
C. Burgess, B. Vis, *Interkosmos*, Springer Praxis Books, DOI 10.1007/978-3-319-24163-0_5

Georgi Ivanov, Bulgarian cosmonaut-researcher. (Photo: Spacefacts.de)

The State Aircraft Factory once existed in Lovech, producing airplanes such as the Laz-7 and Laz-7M, designed by Professor Tsvetan Lazarov. These would fly high over the town and surrounding areas on test flights, attracting the excited interest of many young locals.

As a boy, Ivanov was fascinated by stories of the sea, and would avidly read books about famous ocean explorers as well as adventure stories such as Jules Verne's *20,000 Leagues Under the Sea*. He entertained thoughts of one day becoming a sailor, but found fascination in the mesmerizing sight of the silvery aircraft performing aerobatics overhead and couldn't decide whether his future lay in the ocean or the skies.[1]

By the time Ivanov graduated from the Todor Kirkov technical secondary school in 1958 he had decided to pursue a career in aviation. The decision had been eased by the closing of the Nahimov Navy School in Varna in 1955. In those years, Bulgaria had a large number of model aviation, gliding, and parachuting clubs, plus other forms of flying activities for the enthusiast. From 1955 Ivanov attended many of these clubs, and made his first parachute jump at the age of 16. He joined the DOSO, a junior civil defense organization, and began rigging up high-performance gliders. In 1957 he completed an initial gliding course with the Pleven aero club in the nation's capital of Sofia. Following this he made his first solo flight under the supervision of instructor Minko Alashki. A year later, at the same flying club, he flew a propeller-driven Laz-7 aircraft.

After graduating from secondary school in 1958, Ivanov applied to attend the renowned Georgi Benkovski Higher Air Force School in Dolna Mitropolia, near Pleven. Much to his disappointment his application was temporarily refused on health grounds. Undeterred, he took a course in pilot operation at the flying club in Oryhovitsa. On reapplying for entry to the air force school in 1959 he was successful.

When asked what he was doing when he heard that Yuriy Gagarin had made the world's first space flight, Ivanov recalled the day clearly. "When Gagarin flew, I was a sophomore cadet in the Georgi Benkovski Air School in the [Bulgarian] Republic. On April 12, 1961, it was announced on the radio that a man had flown in space. We had an internship at the Vaptsarov manufacturing [machine building] plant in Pleven, and we heard the message through a speaker, but never knew the name of the cosmonaut. All of us left their jobs, went out into the yard of the plant and began a spontaneous rally. As Air Force cadet students we felt involved in this flight since Gagarin was a military pilot.

"When he took off, and after German Stepanovich Titov, Valeriy Bykovskiy and Valya Tereshkova, we cadets who dreamed of flying airplanes could hardly contemplate even in 2000 that Bulgaria would have a cosmonaut."[2]

Ivanov graduated from the five-year course at the school as a lieutenant in 1964, with a diploma of engineer-pilot and stayed at the school as an instructor, although his ultimate ambition was to become a fighter pilot.

A WHOLE NEW DIRECTION

In 1967 Ivanov gained a promotion to instructor-pilot 2nd class, and the following year was elevated to instructor-pilot 1st class. He transferred to the Combat Section of the Bulgarian People's Air Force in 1968, where he flew a MiG-19 fighter jet and mastered the operation of the MiG-21. That same year he also joined the Bulgarian Communist Party. By this time he had married Natalia Rousanova, who was a public servant, and in 1967 they had a daughter named Ani.

As a later flight commander and squadron leader, he gained the title of military pilot 1st class and received medals for exemplary service to the Air Force. From 1975–1978 he was commander of the 1st Squadron of the 18th Fighter Air Division based at Dobroslavtsi Air Base near Sofia. Under his command the squadron was in the process of transitioning to the MiG-21MH supersonic fighter-interceptor jet aircraft. When he first became aware of the quest to find the first Bulgarian cosmonaut, he held the rank of major and had accumulated just under 2,000 flying hours in eight different aircraft types.

One interesting facet of Ivanov's application was that if selected he would be required to change his Kakalov birth surname. The reason was that in the Russian language it actually resembled a rather obscene word for the function of a person's backside. A compromise was reached whereby he adopted a version of his father's first name, Ivan. From then on he went publicly by the name Georgi Ivanov, which he still uses today.

The criteria used in the initial phase of the 1977 selection process stipulated that all the applicants had to be fighter pilots in the Bulgarian People's Air Force. Furthermore, they must be graduates of the Georgi Benkovski Higher People's Air Force School who finished their studies between 1964 and 1972, which meant that the candidate should have a science degree and at least three years' flying experience.

Contemporary reports state that around 700 officers filed applications. The Aeromedical Commission of the People's Air Force evaluated each candidate, and those who did not pass were sent back to their respective garrisons. As Ivanov recalled in March 2011, "There were many candidates and we were divided [into] groups of ten to twenty people."[3]

The officers who remained qualified were then sent to the Sofia Military Hospital for a complete medical examination that took several weeks. Later, after the rigid qualification standards had drastically reduced the number of candidates, the age limit of the volunteers was raised to allow more officers into the evaluation group. The field was then narrowed to just 15 candidates, ranging in experience from squadron commander to executive officer of an air regiment. Following a new range of evaluations and examinations, that number was further reduced to six, including Georgi Ivanov. The other five candidates were:

- *Aleksandr Panayotov Aleksandrov:* Graduated from the Georgi Benkovski Higher Air Force School as a pilot-engineer in the spring of 1974, subsequently receiving his first commission as a junior fighter-bomber pilot in anti-aircraft units of the Bulgarian Air Force, but entering the evaluations as a civilian candidate.
- *Capt.-Engr. Chavdar Dzhurov:* Said to be the son of Gen. Dobri Dzhurov, the Bulgarian Minister of National Defense. At one time Chavdar Dzhurov held the world record height for a night parachute jump of eight miles. Confusingly, despite official accounts revealing him as one of the 1977 candidates, the Chavdar Dzhurov who was the son of Dobri Dzhurov was killed along with another pilot, Ventsislav Yotov, when their L-29 jets collided in mid-air on 14 June 1972 – some five years before the cosmonaut selection process even began.
- *Capt.-Engr. Ivan Nakov:* Born in 1950, Nakov graduated from the Georgi Benkovski Higher Air Force School in 1972 and subsequently served as an Air Force pilot in the Bulgarian People's Army. After his retirement from the Army he pursued diplomatic work.
- *Lt. Col. Kiril Radev:* 41-year-old supersonic fighter pilot and a squadron commander in the 18th Fighter Aviation Regiment (IAR), which was flying MiG-17, MiG-21 and MiG-23 jet aircraft. He would later claim that even though he was medically fit, he was removed from the list of finalists for political reasons. This probably stemmed from what was deemed a politically unacceptable piece of family history, when it was discovered that his wife had an uncle who had emigrated to Canada after 9 September 1944, in the wake of the Bulgarian Socialist Revolution.

Mystery candidate Chavdar Dzhurov prior to a parachute jump. There is unresolved confusion about the date of his death. (Photo: Chudesa.net).

This extremely minor family irregularity was sufficient on then-existing propaganda grounds to end his candidacy.

* *Maj.-Engr. Georgi Yovchev:* A fighter pilot, Yovchev had graduated from the Georgi Benkovski Higher Air Force School and the N.E. Zhukovskiy Air Force Engineering Academy (Moscow). He became an instructor pilot in the Air Force General Staff of the Bulgarian Defense Force and was the inventor of the UP-9 parachute. He died on 6 May 1983 attempting to land a MiG-19 near Balchik in adverse weather conditions.

Once Dzhurov and Radev had left the group, the remaining four candidates were flown to Moscow, where two more were to be eliminated from the list. Once there, Soviet physicians conducted their own intensive tests and discovered that Yovchev was suffering from a slight cardiological problem, which was sufficient to see him excluded. This left just Aleksandrov, Ivanov, and Nakov. When the final decision was handed down, Ivan Nakov had missed out. He returned to the Bulgarian Air Force.

Aleksandr Aleksandrov. (Photo: Spacefacts.de)

THE MAKING OF A COSMONAUT

In March 1978 the newly renamed Ivanov and Aleksandrov arrived in Star City with their families to commence cosmonaut training. Space flight researcher Maarten Houtman has made the interesting observation that a total of five socialist bloc countries sent their two flight candidates to Moscow at the same time (Bulgaria, Hungary, Cuba, Mongolia and Romania, with the Vietnamese candidates arriving in April 1979) and a clear pattern was established for this second group in that, whether by politically motivated design or by a coincidence, they were launched on missions in strictly alphabetical country order in the Cyrillic alphabet. If the rationale was indeed political, then Romania receiving the final Interkosmos mission was probably a direct result of Moscow's disapproval of Romanian behavior in that country's refusal to participate in the 1968 invasion of Czechoslovakia.

As with the candidates from other Soviet bloc countries, the first stage of training for Ivanov and Aleksandrov included language classes, theoretical studies combined with flight experience in jet aircraft, physical exercises, exercises aboard aircraft which provided brief periods of weightlessness, and performing basic forest and ocean recovery procedures. The second phase, now assigned to a Soviet cosmonaut commander, consisted of mastering the Soyuz and Salyut systems, together with their specific flight program. As a result, prior to commencing the second phase the two Bulgarians were paired with mission commanders. Georgi Ivanov got the twice-flown Nikolay Rukavishnikov (Soyuz-10 and Soyuz-16) and Aleksandr Aleksandrov trained with Yuriy Romanenko, whose only flight to that time was on Soyuz-26 in December 1977, which had linked up with the Salyut-6 space station. Romanenko and flight engineer Georgiy Grechko had set an endurance record by spending 96 days in space.

Rukavishnikov and Ivanov during mission training. (Photo: Spacefacts.de)

Eventually the pairing of Rukavishnikov and Ivanov was formally announced as the prime crew for the Soyuz-33 mission. Rukavishnikov had already gained considerable experience in the Interkosmos program, having served as back-up pilot for Soyuz-28 along with Czech Oldrich Pelcak. In fact, if one looks at the Interkosmos flight list, it is noticeable that from the outset and up until the eighth mission in the series, the back-up commander would wait out the next two flown missions and then be assigned to command the following flight, very much like the American astronaut rotation system. This is interesting in terms of who was appointed during the training period to the role of prime crew commander: Rukavishnikov, Kubasov, Gorbatko, Romanenko, and Dzhanibekov were the back-up commanders for their respective Interkosmos missions, then, without exception, after missing the following two flights in the series they were assigned to command the prime crew of the next Interkosmos mission.

The flight plan for Soyuz-33 was fairly straightforward: to launch, rendezvous and dock with the Salyut-6 space station and spend seven days conducting experiments along with the two cosmonauts then in residence.

The Soyuz-33 rocket travels to the Baykonur launch pad. (Photo: Spacefacts.de)

DOCKING FAILURE

The launch of Soyuz-33 (with the call sign *Saturn*) took place at 8:34 p.m. Moscow time on 10 April 1979, marred by one of the highest wind speeds ever recorded for a Soviet launch, with winds gusting up to nearly 40 miles an hour buffeting the spacecraft during lift-off from Kazakhstan's Baykonur Cosmodrome. As the Soviet news agency TASS reported at the time, "Not one manned spacecraft has blasted off in such bad weather [as that] in which the Soyuz-33 lifted off." Soviet television reported that the launch was nearly postponed.

Nevertheless the ascent went well, and within ten minutes the two cosmonauts were safely in orbit. They had with them vials containing earth from two mountains that had propaganda significance – Buzludja where the Bulgarian Communist Party was born, and Shipka, which served as a symbol of friendship between the two nations.

As recorded by Rex Hall and David Shayler in their 2003 book *Soyuz: A Universal Spacecraft*:

> The crew then spent their first three orbits configuring the Soyuz for orbital flight by deploying appendages, establishing communications with ground- and ocean-based communication stations, and checking the Igla approach and rendezvous system. During the fourth orbit they took off their Sokol suits, opened the connecting hatch to the [orbital module], and prepared a meal, before firing the orientation engines on the fourth and fifth orbits to begin the long flight towards the Salyut. By the next day, during the seventeenth orbit, five orbital corrections had placed Soyuz-33 on the correct approach path to intersect the orbit of Salyut.

Rukavishnikov relayed information to the ground that they were both well and the Soyuz systems were functioning normally. He was instructed to complete the final approach on the next orbit, the eighteenth, and as the two spacecraft closed to within two miles of each other he repeated that everything was proceeding to plan. Vladimir Lyakhov and Valeriy Ryumin, the resident crew of Salyut-6, had been launched aboard Soyuz-32 on 25 February. Then, six hours into their seven-day planned mission, the visiting crew encountered a major problem.

As Hall and Shayler continued:

> It was then that both ground control and the crew noted 'deviations in the regular operating mode of the approach-correcting propulsion unit of the Soyuz 33 engine.' Rukavishnikov later commented that onboard the Soyuz, the crew noted something wrong with the functioning of the engine. It was supposed to fire for six seconds but shut down after only three seconds of erratic firing. Abnormal vibration – felt by both men strapped inside the [descent module] – then shook the Soyuz, and the docking was aborted. The problem was subsequently traced to the gas generator feeding the main engine's turbo-pump.
>
> The crew had only read-outs of the functioning of the engine, and no data to diagnose the fault. Engine firing and shut-down was controlled from the flight control center, with the cosmonauts using a stop-watch and reporting the operation to the ground. This engine would also be used for the de-orbit burn. Despite pleas from the crew to attempt a second approach, flight control wanted to ensure the safety of both the Soyuz-33 crew and the Salyut crew, and decided to follow the mission rules to bring the crew home as soon as possible by using the back-up engine.[4]

In essence, a part that had been tested some 8,000 times on the ground had failed in space, becoming the first orbital engine failure in manned spacecraft history. TASS announced the dramatic circumstances of the flight in a brief dispatch that said, "In the process of approach, there arose deviations from the regular mode of operation of the Soyuz-33 spaceship and the link-up with the Salyut-6 station was cancelled. Cosmonauts Nikolay Rukavishnikov and Georgi Ivanov began preparations for the return to Earth."[5]

The last time the Soviet Union had failed to dock to a manned space station occurred on 10 October 1977, when the Soyuz-25 craft carrying Soviet cosmonauts Vladimir Kovalyonok and Valeriy Ryumin failed to engage the docking latches of the station despite five attempts. Lacking the fuel to maneuver around to attempt a docking at the other end of the station, and with their battery power limited, they had returned home. Since then, all such dockings had proceeded normally, becoming almost routine.

RELUCTANT HOMECOMING

Bitterly disappointed, the Soyuz-33 crew had to put these troubles behind them and prepare for an early re-entry. As there was no immediacy attached to their return, Rukavishnikov and Ivanov carried out some Earth observation exercises before preparing a meal and then settling down for their second night in orbit.

As recalled in April 2001 by flight director for Soyuz-33, Aleksey Yeliseyev, a veteran of Soyuz 5, Soyuz 8 and Soyuz 10, "The docking was impossible and we had to decide how to save the crew. It is hard for me to describe in words what we went through that night. We took a decision that the crew should return using the back-up engine, the third small engine, and all fuel on board, in order to take the spacecraft out of orbit and to start its descent into the atmosphere. We drew many important lessons from that flight, and the cosmonauts' conduct was evaluated as courageous and as the only possible correct conduct."[6]

The two crews had intended to spend 12 April aboard the space station celebrating the 18th anniversary of the pioneering flight of Yuriy Gagarin. Instead, the frustrated crew of Soyuz-33 transferred as much scientific equipment as possible to the descent module, then donned their Sokol suits again and sealed the hatch between the two modules. The orbital module would be jettisoned for the journey home.

With their return home in peril, the two cosmonauts' pre-flight training was critical in conducting the re-entry phase. (Photo: Spacefacts.de)

When they were ready, the two men strapped themselves tightly into their couches as they set procedures in motion for the de-orbit burn. This entailed firing the small back-up re-entry engine on the spacecraft, which was located in close proximity to the faulty main engine and untested If that engine had been breached or badly damaged, igniting it could prove fatal for the crew. Once it had fired, it had to burn for a specific amount of time in order to allow the spacecraft to slip into the extremely narrow re-entry corridor. If it were to burn for less than 90 seconds, the vehicle would be unable to return to Earth in a controlled manner. Should it burn for more than 188 seconds at full power, then the vehicle would re-enter at too steep an angle and burn up. Even if it somehow survived this hellish plunge through the searing heat of the atmosphere, the descent module would be travelling so fast as it approached the ground that the parachutes would likely shred as they were deployed, and the impact with the ground would be fatal.

That day, Mission Control was on full alert. The grim reality facing them was that the chances of recovering the cosmonauts alive were considered slim, and they were mentally preparing themselves for the worst.[7]

Inside the Mission Control Center, Vladimir Shatalov, commander of cosmonaut training, was giving helpful advice to the crew, while at the same time trying to ease their tension with humorous asides. At one stage he advised Ivanov to press down hard on his thick mustache well before pulling down the visor on his helmet.

At the appointed time, Rukavishnikov pressed the button that fired the back-up engine. To everyone's relief it ignited at full power. After three minutes, Rukavishnikov could see that the engine would not stop firing automatically, so he shut it down manually; by which time it had run for an additional 33 seconds. He relayed the news to Mission Control, saying, "The engine burned for 213 seconds … We're coming down on a ballistic trajectory."

Having discarded the propulsive and orbital modules, the Soyuz-33 descent module now hurtled through the ever-thickening atmosphere, compressing air in front of it to form a bow-wave of superheated gases that reached 3,000 °C. Inside their spacecraft, the two cosmonauts could see flames licking past through the portholes, while Rukavishnikov reported hearing a "great noise" and being jolted around.

Eventually, having endured a crushing 9 g's of force on their bodies, the two cosmonauts knew they had survived re-entry. Then the parachutes deployed and billowed open without a problem. The capsule touched down at 7:35 p.m. Moscow time, about 200 miles south-east of Dzhezkazgan in the Karaganda Province of Kazakhstan. They had completed 31 orbits of the Earth and were in space for 47 hours and 1 minute.

By the time that search helicopters finally made it to the landing site, Rukavishnikov and Ivanov were standing beside their spacecraft, which had tipped onto its side. A preliminary medical check indicated that both men were in relatively good condition, with no noticeable after-effects from their ordeal. Georgi Ivanov recalled the dramatic return in a March 2011 interview.

"The main engine of the spaceship installed in the ship's center was damaged. The pump providing fuel to this engine under high pressure stopped working. The back-up engine was also impaired, thus forcing the spaceship to return to the Earth following a so-called ballistic trajectory rather than doing so normally. All spaceships [usually land] smoothly, while we returned along a steep curve. During our descent, the overloading reached 8 to 9 [g's]. That is, if a person weighs 80 kilograms, his or her weight will increase by 8 or 9 times. We were trained to withstand such overloading, but not for more than 90 seconds. Unfortunately, we had to endure it for 2 or 3 minutes. During that time we could neither breathe nor talk. The operators from the Air Route Traffic Control Center (ARTCC) were calling us all the time, 'Saturn, how are you? What is the situation?' Nikolay Rukavishnikov managed only to say, 'Leave us!' In the meantime, through the window, we saw the fire which had wrapped the vessel."[8]

At the 8th Planetary Congress of the Association of Space Explorers, held in Washington, D.C., on 24–29 August 1992, Nikolay Rukavishnikov was asked if it was true Soyuz-33 had landed many hundreds of miles from the target point.

"No, that's not true," was the response from the veteran cosmonaut. "We came quite close to the target point, but that was a pure chance. God helped us … up there. To keep us from perishing He brought us down to that point. It wasn't our own exertions. That landing could have taken place anywhere … even in the Pacific Ocean. The probability of us landing at the calculated spot was like one-hundredth of a percent."[9]

ONE AND ONLY FLIGHT

Following the failure to dock with the Salyut-6 station and a subsequent enquiry into the mishap, there was some speculation that the Soviet Union might attempt a re-launch of the mission, possibly with the back-up crew of Yuriy Romanenko and Aleksandr Aleksandrov. But this would mean that Bulgaria had sent two men into space, much to the chagrin of the other Interkosmos nations. There was some speculation that the original crew would be re-launched, but as time went by it became apparent that this was not going to happen.

A decision was soon reached that many of the Bulgarian experiments that were to have been carried out during Ivanov's seven days aboard Salyut-6 would instead be done by the resident crew of Lyakhov and Ryumin. Accordingly, when the Progress-6 unmanned cargo ship lifted off from Baykonur on 13 May, it carried such items as the Bulgarian Duga and Spektr-15 equipment, the Sredets medical-biological instrument, and the Pirin Metallurgical packet. Progress-6 docked without any problem, and the equipment was duly transferred to the station as part of 100 items of cargo.

Precise details of what happened to cause the life-and-death re-entry of Soyuz-33 would not be revealed until 1986, some seven years after the event. As Velian Pandeliev wrote in his insightful 2012 article, *Bulgaria's First Cosmonaut and the Near-Disaster of Soyuz 33*:

> The mission was subject to all the media restrictions, blackouts and propaganda spinning common to the Soviet space program … When the cosmonauts returned to Earth, the reason for their unsuccessful docking was cited as "a technical glitch", leaving the impression that the docking apparatus had failed, launching a myriad jokes about Soviet technical proficiency but never hinting at the real danger to the cosmonauts' lives.
>
> Not only that, but Soyuz-33 caused a home-grown media gaffe in Bulgaria. On April 11, the day the docking with Salyut-6 was supposed to take place, an issue of the magazine *Bulgarian Warrior* was printed, entirely devoted to the mission.

Originally, two versions of the factual article describing the docking were prepared – one in case of success and one in case of failure. However, an overzealous editor, seeing the success of the launch, approved the printing of the magazine with the successful version in it a few hours too early. The magazine was already on the stands, proclaiming the successful docking and hand-shaking and gift-exchanging of the Soyuz-33 and the Salyut-6 crews when news came through from Moscow that the docking had never taken place. Almost all copies of the magazine were hastily recalled and destroyed, to the point that for decades not even the National Library had a copy of the ill-fated issue. However, individual units survived, squirrelled away here and there, and recently even an almost complete digital scan has come to light.[10]

Post-flight, Georgi Ivanov received the title Hero of the Soviet Union and the Order of Lenin. Back in Bulgaria once again, he was awarded the Hero of the People's Republic of Bulgaria and the Order of Georgi Dimitrov in 1979. Meanwhile, he continued his military service, mastered the MiG-21, served as an inspector in the Bulgarian Air Force and went on to gain the rank of lieutenant general. He defended his PhD thesis in the physical sciences, gained his master's degree in 1984, and then earned a scientific degree as 'Candidate of the Physical Sciences'.

In 1986 Ivanov applied to fly the second Bulgarian space mission because he felt space technology had progressed since his first flight, and though he was a finalist and wanted to participate in its preparation, the flight went to his former back-up, Aleksandr Aleksandrov. He later said he thought it was Aleksandrov's turn to fly and would have been happy with the back-up role. Shortly after passing the first medical tests, he was ordered by the Ministry of Defense to terminate the medical tests "under the pretext that I needed to be preserved as the first Bulgarian cosmonaut". And that's where his participation in the second flight ended.

Ivanov went on to work at the Ministry of Defense, became a member of parliament, and was a member of the Bulgarian Socialist Party. He was elected deputy of the Grand National Assembly that drafted Bulgaria's post-communist constitution in 1991, but when the National Assembly was dissolved that year he was made redundant and pensioned off. Anything to do with his space flight suddenly became part of an unwanted past for Bulgaria. Ivanov said that his name was removed from school books, but many years later it was reinstated in books for 2nd and 3rd grade children.

Asked how he felt about the changes in his country after the collapse of the Soviet Union, Ivanov said that on the positive side, democracy had brought freedom of speech and with it the possibility to realize new dreams. On the negative side, he thought it had led to a brain drain from Bulgaria. He said that he is also worried about so many children and youngsters becoming drug addicts, and the decreasing quality of education. He laments the decline of the Bulgarian space program, but is nevertheless optimistic about the future of world space programs.

In 1993 Ivanov said he accidentally ended up in the aviation industry, thanks to a good friend, and is now the executive director of Air Sofia Ltd., a private transportation company. In another commercial venture in 2000 he gained ownership of the new 18-hole Ihtiman golf course and hotel complex. After his marriage to Natalia ended in divorce in 1982 he married again, this time to Lidia, an economist with a PhD and they have a son, Ivan, born in 1984.

Georgi Ivanov and Aleksandr Aleksandrov at a function in Moscow's Kremlin Palace on 12 April 2011 to celebrate the 50th anniversary of the flight of Yuriy Gagarin. (Photo: Bulgarian Cultural Institute)

On 12 April 2011, by Presidential Decree No. 435, Georgi Ivanov was awarded the medal "For merits in space exploration".

Back-up pilot Aleksandr Aleksandrov flew his space mission aboard Soyuz TM-5, which launched on 7 June 1988 and succeeded in docking with the Mir Space station. He went on to become deputy director of the Institute of Space Research at the Bulgarian Academy of Sciences. Both men still make occasional public appearances.

Today the Soyuz-33 capsule is housed in the Plovdiv Museum of Air in central Bulgaria, along with much of Ivanov's paraphernalia from the aborted mission.

Nikolay Rukavishnikov, the commander of Soyuz-33, had an unhappy future in regard to the occupancy of a space station. He flew on Soyuz-10 in 1971. This was the first manned flight to a Soviet space station, but a malfunction prevented proper docking with Salyut-1. He then flew a six-day Soyuz mission in 1974 as a precursor to the Soviet-American ASTP mission. On the Soyuz-33 mission in 1979 he and Bulgarian Georgi Ivanov failed to dock with Salyut-6. He was scheduled to fly in 1988, this time to the Salyut-7 station, but had to be replaced by Aleksandr Serebrov owing to illness. On four orbital flights he would never enter a Soviet space station.

REFERENCES

1. Borislav Evlogiev, article "1979: A space odyssey," *Bulgarian Air* magazine, Spring 2009. Website: *https://www.air.bg/content/magazine/pdf/63.pdf*
2. Eliana Mitova interview with Georgi Ivanov for klassa.bg.news article, "Georgi Ivan: the first Bulgarian astronaut," Sofia, Bulgaria, 31 March 2011. Website at: *http://www. klassa.bg/News/Read/article/164058*

3. *Ibid*
4. Rex Hall and David Shayler, *Soyuz: A Universal Spacecraft*, Praxis Publications, Chichester, U.K., 2003
5. *Sydney Morning Herald* newspaper (Australia), unaccredited article "Space link-up fails, cosmonauts return," issue Friday, 13 April 1979
6. Borislav Evlogiev, article "1979: A space odyssey," *Bulgarian Air* magazine, Spring 2009. Website: *https://www.air.bg/content/magazine/pdf/63.pdf*
7. Velian Pandeliev, article, "Bulgaria's First Cosmonaut and the Near-Disaster of Soyuz 33, *Blazing Bulgaria* blog, 14 June 2012. Website: *http://blazingbulgaria.wordpress. com/2012/06/14/soyuz-33*
8. Eliana Mitova interview with Georgi Ivanov for klassa.bg.news article, "Georgi Ivan: the first Bulgarian astronaut," Sofia, Bulgaria, 31 March 2011. Website at: *http://www. klassa.bg/News/Read/article/164058*
9. Bert Vis interview with Nikolay Rukavishnikov, Congress of the Association of Space Explorers, Washington, DC, 24–29 August 1992
10. Velian Pandeliev, article, "Bulgaria's First Cosmonaut and the Near-Disaster of Soyuz 33, *Blazing Bulgaria* blog, 14 June 2012. Website: *http://blazingbulgaria.wordpress. com/2012/06/14/soyuz-33*

6

Hungary joins the space club

Collectively, they are among the bravest and most level-headed groups of men on the planet. Yet Russia's cosmonauts have always been a superstitious lot. Prior to embarking on their flights into space they ritually observe traditions, many relating to the memory and legacy of Yuriy Gagarin. This extends to visiting the Memorial Wall in Red Square to leave carnations where Gagarin and four cosmonauts who died on space missions are interred; they visit his office (maintained exactly as it was on the day that he died) and it is said they ask his ghost for permission to make their flight. They will not watch the roll-out of their carrier rocket, as this is considered bad luck; instead, they each have a haircut.

The evening before their mission they sit through a viewing of the cult 1969 movie *White Sun of the Desert*, and as they leave the Cosmonaut Hotel on launch day – it would be considered bad luck to stay elsewhere – it is to the tune of *The Grass Near My Home* by the band Zemlyane (Earthlings). On the way out to the launch pad their transfer bus stops so the male cosmonauts can urinate against the right rear wheel. Obviously female cosmonauts cannot perform this long-standing superstition; instead, they are permitted to carry a small container of their urine and tip it onto the wheel if they wish. Also, many will not sign autographs until after they have been into space the first time.

The selection of Bertalan Farkas as the first Hungarian research cosmonaut caused several of the cosmonauts to complain about potential bad luck associated with his upcoming flight aboard Soyuz-36. The reason was his mustache. On the previous Interkosmos mission, that of Georgi Ivanov on Soyuz-33, the docking with Salyut-6 had had to be abandoned. Ivanov was sporting a thick black mustache. Consequently, amazing as it might seem, the fact that anyone would dare to wear a mustache into space was seen as a bad omen by the cosmonauts. It was therefore recommended to Farkas that he shave off his own mustache, but he refused. Had any grievous incident happened on his flight, then it is doubtful that any future Russian cosmonauts would have been launched unless they were clean-shaven.

© Springer International Publishing Switzerland 2016
C. Burgess, B. Vis, *Interkosmos*, Springer Praxis Books, DOI 10.1007/978-3-319-24163-0_6

A SHOEMAKER'S SON

Bertalan Farkas (known to many as Berci) was born on 2 August 1949 in the village of Gyulahaza, situated in the Szabolcs-Szatmar region of north-east Hungary. His father Lajos had left a tradition of farming the land to become a skilled master shoemaker in the nearby Kisvarda Shoe Industry Cooperative, while his mother Erzhebet took care of the family.

After completing grammar school Berci was enrolled in the Georgi Bessenyei Vocational High School in Kisvarda. There he soon developed a strong passion for football, and local coaches saw potential in the young man. Some even suggested he should seriously consider a career in the sport, but life would point him in a vastly different direction. "When I was a child I had no thoughts of becoming a spaceman," he recalled. "I seem to remember I wanted to be a footballer. I was a big fan of Ferencvaros."[1] However his parents considered football too dangerous, so he reluctantly set those plans aside.

In his later high school days Berci had decided to pursue a career as a mechanic, but one of his classmates, Gyuszi Kopasz, suggested that they join a flying club and learn to become pilots. Berci had very little interest in aviation, but his friend persuaded him and they both decided to apply to a flying club attached to the Hungarian National Defense Association in Nyiregyhaza, a junior defense organization. In a twist of fate, whereas Berci easily passed the preliminary medical examination, Gyuszi failed due to high blood pressure and was not admitted to the course. Through the flying school, Farkas learned to parachute jump and then became a glider pilot.

Friends as youths and later cosmonaut candidates Bertalan Farkas (left) and Bela Magyari as first class fighter pilots in the Hungarian People's Army. (Photo: Author's collection)

In the spring of 1965, while still attending the Georgi Bessenyei High School, Farkas also befriended a fellow flying enthusiast named Bela Magyari, who had been born six days after himself. Magyari had a yearning to learn to fly in motorized aircraft, a prospect which also interested Farkas. They both studied in single-engine trainers at Nyireghaza airport under instructor Gyula Rozman, and after passing the demanding six-week course they decided to apply for pilot officers' school once they had graduated from high school.

In 1967 Farkas passed the matriculation examination and enrolled at the Kilian Gyorgy Flight Technical College in Szolnok, Central Hungary, to study aeronautical engineering. His mother argued that she did not want him to be involved in such a dangerous occupation, but eventually he signed up against her wishes.

At the Kilian college he found the work difficult and demanding, but after two years he was determined to become a fighter pilot and made a decision which would change his life forever – he was about to continue his education at the Krasnodar Military Aviation Institute in the Soviet Union. Accompanying him in this new adventure was his good friend Bela Magyari. In 1972 both men graduated as pilot officers and moved back to Hungary, where they joined the Hungarian People's Army and were later commissioned as 2nd lieutenants in the Hungarian Air Force. Farkas was then assigned to a fighter pilot squadron in a national defense fighter unit. It was while stationed there that he would meet his future wife, Aniko Stuban, a photographer. They were married in 1974. He then served as an instructor with several fighter-interceptor units and in 1977 qualified as a fighter pilot 1st class.

"By 1978 I had flown more than 1,000 hours as a fighter pilot," he stated, "flying faster than the speed of sound many times."[2]

OVER TO MOSCOW

Bela Magyari was born on 8 August 1949 in the city of Kiskunfelegyhaza, Bacs-Kiscun County, situated 80 miles south-east of the Hungarian capital, Budapest. He attended local primary and secondary schools in his hometown, later becoming active in the Communist Youth Union. At school he was known to be a highly motivated student who excelled in mathematics and physics, and would even organize study competitions in which he was himself very successful. At this early stage of his life he had no real interest in aviation.

When Bela was 15 years old he began failing quite badly at mathematics, preferring instead to spend a lot of his normal study time playing basketball. Soon enough, however, things would quickly change. His father, who worked as the school's janitor, noted the slide in his education and encouraged his son to devote more attention to his studies and also to take up the healthy activity of cycling. The two of them would generally ride together. At weekends they would make their way out to the local airport, where several of his relatives were employed, and Bela would spend a pleasant hour or two just watching airplanes taking off and landing. One day his uncle Aleksey asked Bela if he would like to join him on a short flight, and he could not believe his luck. Within minutes of strapping in, they were airborne. "The flight was only five minutes," Magyari recalled, "but what a wonderful five minutes." From that time on he knew where his life would take him – into the skies.

His youthful enthusiasm fired, young Bela began hanging around the airport, doing odd jobs and washing airplanes over the summer months. He soon began asking about learning to fly, and was encouraged to do so by his uncle, but first he needed to learn the theory of flight. He was presented with "three or four thick books" on the subject, and informed that he would be examined on their contents in three weeks. He was still only 15 years of age so he had to get the consent of his parents to continue in his dream of flying, but to his joy they gave him permission.

Everything went to plan, and Bela passed the technical examination, following which he was given practical training. Eventually it was time to do a solo flight. "I was a little awed; it was an incredible feeling to be up in the air," he reflected. "But I was not completely relaxed. One moment I would be filled with joy, and would shout with happiness, but at other times [controlling the aircraft] was difficult – a real pain in the butt. I'll never forget that day. Then I decided that, if possible, I would like to make a living in aviation."[3]

In 1966 he heard about a great opportunity to further this interest at the Kilian Gyorgy Flight Technical College in Szolnok. After completing secondary school the following year he applied and was accepted at the college. He graduated in 1969 and then continued his studies at the Krasnodar Military Aviation Institute, near the Black Sea in south-western Russia.

Magyari was promoted to flight lieutenant in 1972 and assigned to the Hungarian Air Force. The following year he joined the Hungarian Socialist Workers Party, and became a fighter pilot 3rd class. By 1977 he had advanced to fighter pilot 1st class. A year later, he became a squadron leader and was promoted to captain. That same year, 1978, he began training as a candidate for the Soviet-Hungarian joint space mission.

AND THEN THERE WERE TWO

The specific criteria for possible selection as the Hungarian cosmonaut-researcher included a height not exceeding 5 feet 10 inches in order to fit into the Soyuz capsule, and an upper age limit of 35. Any candidate had to have a proven proficiency in flying and navigating. They also had to possess an aptitude for learning the Russian language. The respected Hungarian physician Col. Dr. Jozef Szabó and representatives of the Ministry of Defense made a tour of military units which were equipped with the MiG-21, to undertake an initial assessment of potential candidates. The initial screening threw up the names of around 1,000 pilots, but a check of their flight experience and service records cut this number to 206. A basic medical screening reduced the number even further to 110, and then a far more comprehensive set of medical tests was imposed on the candidates. 59 were rejected due to a number of medical ailments; others needed some form of surgery; seven needed dental treatment, and six had neurological disorders. After several rounds of examinations 35 candidates remained. Of these, eight decided not to proceed once further details had been outlined to them, leaving just 27 to complete further tests. That number included friends Bertalan Farkas and Bela Magyari.

In September 1977 the remaining candidates were subjected to an intense two weeks of scrutiny involving 100 individual tests at the Scientific Institute for Aviation in the central Hungarian city of Kecskeme. At the briefing for these pilots, it was noticed that one person was missing. He finally turned up ten minutes later with apologies as his wife had just given birth to their first child. Instead of receiving a rebuke for his tardiness the whole group burst into a round of applause. At the end of the fortnight, their number was down to just eleven.

By December, the results of all the tests and surveys had eliminated another four. Those now remaining were:

- Imre Buczko
- Laszlo Elek

- Bertalan Farkas
- Peter Gutyina
- Have Neumann
- Bela Magyari
- Endre Weigel.

In the next step of the selection process, a team of medical specialists arrived from Moscow to conduct their own evaluations of the men. As recorded in the 2007 Springer-Praxis book *Russia's Cosmonauts: Inside the Yuri Gagarin Training Center*, the range of tests included:

"will power, intelligence, temperament, character, self-control and physical endurance. Two of the leading Hungarian aviation medicine experts were also involved in the selection process; the medical tests were supervised by physician Colonel Dr. János Hideg, and the psychological tests were led by Colonel Dr. Jozef Szabó. As a result of these tests, four finalists, all serving captains in the Hungarian Air Force, were selected at the end of 1977. They were: Imre Buczko, Laszlo Elek, Bertalan Farkas and Bela Magyari."[4]

Once Soviet officials were satisfied with the credentials of the four finalists they were sent to the Gagarin Cosmonaut Training Center in Star City on 21 January 1978 for five weeks of evaluation. Knowing it might be pivotal in his chance of selection, Farkas gave up smoking, which he said was one of the toughest disciplines of all. Additionally he was harboring some doubts about what he was trying to achieve.

"Star City was the hardest for me," he said. "After two weeks, when there are only four of us, came a moment when I said, 'Thank you, I do not want to be a cosmonaut.' But a doctor friend of mine convinced me to do more. I stayed, I persevered, I went through, and then it was the result."[5]

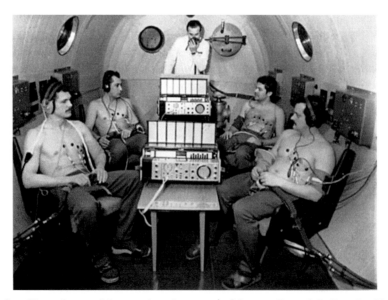

The four Hungarian candidates undergoing tests in Moscow. From left: Bertalan Farkas, Laszlo Elek, Imre Buczko and Bela Magyari. (Photo: Author's collection)

Flanked by white-coated technicians are (from left): Imre Buczko, Bertalan Farkas, Bela Magyari and Laszlo Elek. (Photo: Author's collection)

Farkas and Magyari were informed that they were the two pilots who would train for space and represent Hungary on the international flight. Both men had young families, and on their arrival in Star City they were met at a brand new apartment complex by Maj. Gen. Aleksey Leonov, who mischievously placed two door keys behind his back and allowed the women to choose which apartment they would occupy for the next two years.

Farkas with his wife Aniko overlooking Budapest and the Danube River. (Photo: Author's collection)

COSMONAUT TRAINING

Soon enough, it was time for the two men to knuckle down to training. "I thought it would be easier," Farkas later said of the fatiguing months of studies, language classes, physical training, spacecraft familiarization, and flight preparation. Their daily schedule was quite rigid: by 9:00 a.m. they were already in the classroom to study ballistics, astronomy, and the basic operating elements of spacecraft and space stations, and their day would normally run to 7:00 p.m. There was a lot of complex material to be studied as they also grappled with the Russian language, and constant practice. If there was a problem in what they were studying, the training would often extend to 11:00 p.m.

Bertalan Farkas was eventually paired with Soviet cosmonaut Valeriy Kubasov, a veteran of the Soyuz-6 mission in which he had conducted the first welding experiment in space, as well as the Apollo-Soyuz Test Program (ASTP) mission in July 1975, in which a two-man Soyuz crew of Kubasov and Aleksey Leonov had docked in orbit with an Apollo spacecraft carrying three U.S. astronauts. Kubasov, as Farkas knew, was married with two children.

The Kubasov and Farkas families get together on a social visit. (Photo: Author's collection)

"I already knew him and he often visited our courses," Farkas later revealed, "but I hardly dared think that he would sometime be my commander. He is not only very knowledgeable and a very experienced expert, but he is also unusually friendly. At first I was very afraid of him … How could I approach him? He had twice been in space and his head was full of knowledge. What could this [Interkosmos] flight teach him that he didn't already know? And more importantly, how could I give him any assistance in his work? Later it became clear that Valeriy did not feel this difference at all. We became accustomed to him providing us a great deal of aid in our preparation whenever any kind of problem arose. Kubasov's approach was different. He decided what had to be learned and what we had to be aware of. He assigned a task, and there it was – do it yourself. He did not expound ahead of time. Questions could be asked, but only after we had prepared the material. He was convinced that I had thoroughly succeeded in mastering the material and that I knew a great deal about solutions, and then he would explain what I still had to do and what it would be best for me to study and practice.

"He helped me like an elder brother and a real friend. I really have a great deal to thank him for."[6]

Magyari photographed with Soyuz-36 cosmonaut Dzhanibekov. (Photo: Spacefacts.de)

Meanwhile Bela Magyari was teamed up with Vladimir Dzhanibekov, who had recently returned from his first space flight as commander of the Soyuz-27 mission, along with flight engineer Oleg Makarov. Launched on 10 January 1978, they had docked with the Salyut-6 space station, and five days later had returned home in Soyuz-26. This change-over of craft was carried out as a routine safety measure, because an extended period of exposure to raw space could lead to the degradation of a craft's engine, propellant seals, and batteries.

Once the first phase of their training had come to an end, the two Hungarian candidates moved on to the next stage, which encompassed practice flights, parachute jumps, parabolic flights in a specially converted aircraft to experience brief periods of weightlessness, basic familiarization with the Soyuz spacecraft and Salyut space station, long hours spent on the centrifuge, jungle training and water splashdown survival techniques.

Kubasov and Farkas training with the Salyut station mock-up in Star City. (Photo: Author's collection)

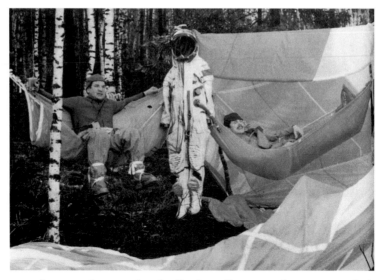

In the event of an anomalous parachute landing into a forest area, the cosmonauts had to train in wilderness survival techniques. (Photo: Mora Konyvkiado)

Farkas revealed that, for him, the toughest part of the training regime were the exercises concerned with marine survival, in which the crews simulated landing on water instead of dry land. He found this to be a very burdensome exercise. They next moved on to spending time in the simulators, exact models of the Soyuz and Salyut crafts. Initially this was an exercise in familiarization with the different systems, but it soon advanced to the proper handling of the onboard equipment and instruments. Finally, they had to solve complex simulated flight tasks and problems thrown at them by the training staff. One of these involved the Soviet commander losing consciousness shortly after lift-off, leaving the Interkosmos candidate to carry out complex and unusual tasks by himself in order to save the mission and their lives.

There were spacesuit fittings and exercises, and one humorous occasion on which Farkas and Magyari spent a short time in a tub filled with plaster with a dozen people looking on. "Although it was not the most comfortable moment, I do recall the time spent in the plaster tub," Farkas explained. "An exact plaster model was taken of the back of our bodies. In this way the spaceship seat was fitted to our measurements with unusual precision, helping us to bear the extra pressure when ascending and descending – the pressure exerted on our chest and back – because the back must conform exactly to the seat."[7]

Recreation time generally comprised of physical education and sometimes a little soccer, but the training staff was very wary of overly strenuous activities that could result in broken limbs or other serious injuries. Therefore, the two men would often enjoy a stroll in the space center's park or the neighboring forest.

A collective photo of a number of Interkosmos crewmembers along with cosmonaut dignitaries from Star City. Front row: Pham Tuan, Viktor Gorbatko, Vladimir Remek, Aleksey Gubarev, Dumitru Prunariu, Pyotr Klimuk, Aleksey Leonov, Miroslaw Hermaszewski, Arnaldo Tamayo Mendez, Anatoliy Berezovoy. Back row: Sigmund Jähn, Vladimir Dzhanibekov, Georgi Ivanov, Yuriy Malyshev, Rakesh Sharma, Bertalan Farkas, Leonid Popov, Boris Volynov and Valeriy Bykovskiy. (Photo: Author's collection)

ON TO BAYKONUR

In the wake of the failure of Soyuz-33 to dock with Salyut-6, changes had to be made to the Soyuz flight manifest. Initially, one can assume that Kubasov and Farkas would have flown the Soyuz-34 mission to the space station and, if successful, returned in Soyuz-33 to leave the fresher spacecraft for the resident crew. Using earlier flights to the station as a precedent, it is likely that if Soyuz-33 had completed the docking as planned, Kubasov and Farkas would have been launched about 80 days later, since the operational life of a Soyuz at the time was around 90 to 100 days. Therefore because Soyuz-33 launched on 10 April 1979, this would have meant launching Soyuz-34 around 29 June. But after the Soyuz-33 failure, the Soyuz engine was re-engineered. As the Soyuz-32 craft already docked to Salyut-6 had the same suspect engine as the one on Soyuz-33, it was decided to launch Soyuz-34 unmanned on 6 June to supply the resident crew of Vladimir Lyakhov and Valeriy Ryumin with a reliable return vehicle. The current Salyut residents, Leonid Popov and Valeriy Ryumin, had been launched aboard Soyuz-35 in April 1980. Hence, Kubasov and Farkas were re-allocated to the Soyuz-36 mission.

With their training finally at an end, the two crews made their way to Baykonur for final preparations. This is described in the Hungarian report of 26 May 1980, *Magyar Urhajos a Vilagrurben, MTIjelenti*, which basically translates to *Hungarian Cosmonaut in Space*.

> The space foursome said goodbye to the unaccustomed cold of Moscow, and found every tulip at the height of its flowering back in Baykonur. The former plain was covered with an enormous complex, and the spaceport was waiting for the members of the new international expedition with feelings of warmth.
>
> The last days in [Star City] were unusually busy. In the last examination, scarcely two weeks before launch, Berci Farkas and Bela Magyari received equally outstanding reports, this time as earlier. In the course of two years of difficult work there was no other type of examination report, and in the well prepared space unit this is considered an outstanding achievement. The final medical examination also had a unanimous result: both candidates were fit to carry out the planned program of the expedition, to fly in space. Then they said farewell to their family and friends, and took off for Baykonur.
>
> The Hungarian space candidates had already walked over the spaceport and were familiar with this enormous space complex, but this trip was different. It was now necessary to put an end to the two years of difficult work, and their journey was now to be into space. In accord with the newer preparatory program, the two had to undergo their final preparation at the spaceport and had to reach the launch site earlier than their predecessors. Ten difficult, work-filled days awaited these young men, with their red, white and green badges, and their Soviet commanders. Here with the aid of experts they were molded into final form, determining almost by the second the detailed program of the space trip, and here they had to prepare the complete on-board documentation.

The next most important step in the launch process was the confirmation of the prime and back-up crews. Although Farkas hoped he would be selected, he also felt sad that he was up against his old friend for the only available Soyuz seat. "You know that the two of us cannot fly, that one of us must remain here," he reflected at the time. "Naturally, I want to be the one who is successful, but if Bela is chosen I will be almost as happy."[8]

Eventually the chairman of the State Commission made the announcement, saying that it had not been an easy decision because both teams were equally good. Kubasov and Farkas were the prime crew, with Dzhanibekov and Magyari in the back-up role.

As a member of the commission later observed, "I would have been very happy to launch both pairs on the voyage. Unfortunately there is no chance of doing this now, and so only one pair can go."

HEADING FOR SALYUT-6

On 26 May 1980, Bertalan Farkas became the first Hungarian to fly into space. It would be some 27 years before another would do so and on that occasion, in April 2007, space flight participant Charles Simonyi personally paid for a seat on Soyuz TMA-10 in order to visit the International Space Station; in fact, Simonyi went on to make a second paid flight to the ISS in March 2009 aboard Soyuz TMA-14.

Lift-off for the Soyuz-36 mission. (Photo: Spacefacts.de)

Soyuz-36, the fifth Interkosmos mission, lifted off with Valeriy Kubasov and Farkas from Baykonur at 7:20 p.m. Moscow time with the call sign *Orion*. A docking with Salyut-6 was successfully completed without incident the following day, and several hours later Kubasov and Farkas were warmly welcomed into the station by Leonid Popov and Valeriy Ryumin. Their visit had been preceded a month earlier by the unmanned automatic Progress-9 craft, which had carried around 2,900 pounds of cargo of food, 180 liters of water, clothing, dust collectors, regeneration equipment, spare parts, tools, scientific instruments, film, and mail, plus some scientific instruments to be used by the visiting crew of Soyuz-36. It also brought up a new motor for the Biogravstat centrifuge and gas filters for the station's control system.

Ahead of the arrival of Soyuz-36, Popov and Ryumin had completed the unloading of the Progress-9 cargo craft and then filled it with rubbish and other equipment that was no longer required. It was then remotely undocked on 20 May in order to free up the docking port for the Interkosmos crew. After a period of independent flight, the cargo ship was directed into the atmosphere, where it burned up during re-entry.[9]

Once the visitors had settled in, they set to work to fulfill a research program drawn up by Hungarian scientists. Experiments would be conducted in several fields. Principal amongst these was Biosphere-M. With the 'M' standing for Magyar (Hungary), this was a study of natural phenomena by visual observations, divided into four basic categories. The first was called the Carpathian, which involved taking photographs of geomorphological sites on the ground, including the Carpathian Mountains and areas of archaeological and anthropological value such as Lake Titicaca in South America. The Metamorphisis study meant locating and recording on film geological contours and fracture lines in the Earth's crust, as well as taking pictures of volcanoes and new river channels. Additional Biosphere-M studies were carried out of oceanic and meteorological phenomena. Three areas of Hungary were of particular interest, and this research involved simultaneous studies from space and by aircraft on the same ground track. Kubasov and Farkas photographed over 60 percent of the total area of Hungary using the MKF-6M camera while at the same time an Antonov AN-30 airborne laboratory documented the same area from altitudes of four to five miles. In addition, four types of film were taken from helicopters flying at altitudes between 5,000 and 8,000 feet. Heat-sensing devices also took measurements.

Other experiments included the Diagnosticator, which measured various physiological parameters in the cosmonauts to obtain an objective picture of their state of health before, during, and after the flight.

The Interkosmos crew were also to obtain new data by technological experiments using the Kristall and Splav furnaces. Aimed at the further investigation of melting, diffusion and crystallization in conditions of weightlessness, these experiments were considered to be of great importance for improving the processes for series production of new semiconducting materials and for the further development of electronics.

In accordance with plans, Kubasov and Farkas were to exchange their spacecraft for the previously docked Soyuz-35 on the forward port of the station in order to leave the fresher Soyuz-36 for the later return of Popov and Ryumin. This meant that the four cosmonauts had to transfer their couches, spacesuits and personal effects to the other craft. This process also involved centering weights inside each of the two Soyuz vehicles to ensure a proper center of mass for the returning craft so it did not overshoot or undershoot the targeted landing area.

Popov and Ryumin working with Farkas aboard Salyut-6. (Photo: Spacefacts.de)

By 3 June the international crew had wrapped up their six-day work program, and after making their farewells to the resident crew the two men boarded Soyuz-35. At 12:50 p.m. they successfully undocked and slowly withdrew from the station. A little over three hours later, at 4:07 p.m., the Soyuz-35 descent module safely touched down some 88 miles south-east of Dzhezkazgan. The two cosmonauts had spent a total of 188 hours and 46 minutes in space.

The cosmonauts sign autographs following the safe touchdown. (Photo: Author's collection)

When helicopters from the tracking group flew into the landing site they found the two cosmonauts already unloading their scientific material. A preliminary medical examination reported them to be in excellent condition.

RETURN TO HUNGARY

At a Kremlin ceremony on 10 June 1980, Leonid Brezhnev, General Secretary of the Communist Party of the Soviet Union Central Committee and President of the Supreme Soviet of the Union of Soviet Socialist Republics, presented awards to the Soyuz-36 crew. By decree of the Presidium of the USSR Supreme Soviet, Valeriy Kubasov, already twice Hero of the Soviet Union, received the Order of Lenin, while Bertalan Farkas was awarded the title of Hero of the Soviet Union and the Order of Lenin with the Gold Star medal. In awarding these presentations, Brezhnev said, "We are witnesses to the history of space exploration. Man is increasingly deeply and fundamentally mastering the intricate art of living and working outside our planet. And we are legitimately proud that citizens of socialist countries have contributed great services to mankind in this peaceful realm as well."[10]

Kubasov accompanied Farkas on his triumphant return to Hungary. (Photo: Author's collection)

Farkas became a brigadier general on 22 May 1995. Between 1996 and 1997 he was Hungary's military attaché at the Hungarian Embassy in Washington, D.C. On 31 October 1996, three months after his appointment, he was stopped by police in Virginia and charged with driving while intoxicated. He protested that even though he had been to a reception he had only consumed a single glass of wine. After the Hungarian Ministry of Foreign

Affairs launched an investigation into the charge, Farkas "at his own request" resigned from the post. U.S. authorities asked the embassy for a waiver of diplomatic immunity, but that was denied in December. The following month, he was recalled to Budapest where the Defense Ministry set up a panel to consider disciplinary action. What action was taken was not revealed, but the fact that he retired from the air force that same year seems too much of a coincidence not to be a direct consequence of the incident.

Bertalan Farkas photographed with Valeriy Kubasov, who passed away in February 2014. (Photo: Unaccredited)

In April 1998, Associated Press reported that Farkas had returned to the United States to work at the U.S. headquarters of Orion 1980, a venture he had jointly founded with former shuttle astronaut Jon McBride to promote the manufacture of space-related technologies.

He became a member of the Hungarian Democratic Forum, a Hungarian conservative political party, and was its candidate at the 2006 parliamentary election for the district of Baktalórántháza. He is currently a Commander (CLJ) of the Military and Hospitaller Order of Saint Lazarus of Jerusalem in Hungary, and is president of Airlines Service and Trade.

In December 2014 Péter Szijjártó, Minister of Foreign Trade and Affairs, asked Bertalan Farkas to become his ministerial advisor, with the appointment announced in the Hungarian Press Agency.

Bertalan and Aniko Farkas have three children: a daughter Aida, born 1977, and two sons Bertalan born 1981 and Adam born in 1983.

For his part, Bela Magyari graduated from the Budapest Technical University in 1986. He then spent the remainder of his active career as an aeronautical engineer and pilot. He served as president of the Hungarian Astronautical Society and also worked for the Hungarian Space Office. Married to Marthja and with three children, he is now retired. Today, Bertalan Farkas and Bela Magyari still remain close friends.

A recent photo of Bela Magyari (right) with Hungarian-born space flight participant Charles Simonyi, who has flown twice to the International Space Station aboard Soyuz TMA-10 in 2007 and Soyuz TMA-14 in 2009. (Photo: Veres Viktor)

The Soyuz-36 capsule which carried Farkas into space (and carried a different crew back home) is currently housed in the Vietnamese People's Air Force Museum in Hanoi, Vietnam, while the Soyuz-35 descent module which brought him back to Earth is in the Kozlekedesi Muzeum in Budapest, Hungary.

REFERENCES

1. Budapest Pocket Guide, unaccredited interview with Bertalan Farkas, available at: *http://www.budpocketguide.com/TouristInfo/famous/Famous_Hungarians03.php*

2. *Ibid*
3. Gyongyosi Balasz, article, "Interview with Bela Magyari" (translated from Hungarian), *Szeged-Delmagy* newspaper, 16 September 2009
4. Rex Hall, David Shayler and Bert Vis, *Russia's Cosmonauts: Inside the Yuri Gagarin Training Center*, Praxis Publishing, Chichester, U.K., 2005
5. Victor Harta, *Pecistop* interview with Bertalan Farkas, available online at: *http://peci-stop.hu/tudomany/koszonom-en-akarok-urhajos-lenni-interju-farkas-bertalannal/1241693*
6. Translation of *Magyar Urhajos a Vilagrurben MITjelenti* (basically *Hungarian Cosmonaut in Space*), 26 May 1980, from NASA Technical Memorandum (NASA TM-76719), NASA Washington, D.C., September 1981
7. *Ibid*
8. *Ibid*
9. *Soviet Weekly*, unaccredited article, "Crew Prepares for New Arrivals," 31 May 1980, pg. 3
10. *Soviet Weekly* unaccredited newspaper article, "Space Crew Receives Distinctions for Excellent Flight," issue 24, June 1980

7

A Vietnamese cosmonaut

In 1980 Pham Tuan, a 33-year-old native of Vietnam, became the first person from a Third World country to fly in space. At the time, he was regarded by Vietnam's hard-line communist leaders as the ideal candidate to become that war-torn nation's first man to fly into space. From humble beginnings, he had already risen to the rank of national hero during sustained American air attacks in the Christmas Bombings of 1972, when he was credited with becoming the first Vietnamese fighter pilot to shoot down a B-52 Stratofortress in air-to-air combat. It was a feat many U.S. aviators still regard as highly improbable.

BORN INTO POVERTY

Pham Tuan was born into a peasant family on 14 February 1947. His family lived in the Quoc Tuan Commune on the banks of the Tra Ly River, located in the Kien Xuong District of the province of Thai Binh. This province lies within the northern coastal area of the Socialist Republic of Vietnam and is named after the Vietnamese words for Pacific Ocean. He was one of three children, but the names of his parents have never been released in the media.

His impoverished background was the stuff the Soviet propagandists' dreams were made of; he became only the second child from his village to go to secondary school, according to an article in the *Vietnam News* in 2000. "Even as a young boy he often dreamt of flying as he worked with his siblings herding buffalo and working in the fields," the newspaper said.

"Among his greatest joys then was to watch kites soar … he still preserves the kites that his mother gave him."[1]

Said to have been an exceptionally bright student in high school, he was admitted into the Vietnamese Communist Youth Union in 1963 after completing his 10th grade studies. When the U.S. Air Force began carpet bombing over North Vietnam for the first time in 1964 he was 17 years old. Reacting to this, Tuan and many of his friends volunteered to join the Vietnamese People's Army and he was accepted as a mechanic in July 1965. His original plan was to join the navy, but as one of the few young volunteers with a high-school education he was soon reassigned to the Vietnamese Air Force to train as a pilot. "When I met a military officer, I proposed that he send me to a unit in the air force," Pham Tuan recalled in an interview, "because I felt a great hurt when seeing the helpless eyes of Vietnamese people under U.S. bombings and no one could do anything to stop them."[2]

© Springer International Publishing Switzerland 2016
C. Burgess, B. Vis, *Interkosmos*, Springer Praxis Books, DOI 10.1007/978-3-319-24163-0_7

To his great disappointment Tuan then encountered a problem when his initial medical examination found that his heart beat faster than normal and that he had eyesight problems caused by drinking polluted water. As a consequence, he said, "I was selected to be trained in the former Soviet Union in a maintenance personnel role for the air force." With the war intensifying, however, Vietnam urgently asked the former Soviet Union to select more pilots among the maintenance trainees. As a result, Tuan was pressed into flight training in 1966 due to this severe shortage of pilots.

As reported by space historian Gordon Hooper, "This delay caused him to arrive at the Soviet combat pilots' school three months late, when his colleagues had already completed their theoretical training. However, in all the theoretical subjects Tuan reportedly got the highest marks. He did well in both day and night flying and graduated in May 1968 with a pilot's certificate which stated that he was a 'strong willed and very promising' pilot."[3]

Following his graduation, Tuan returned home to Vietnam and that November became a member of the Communist Party. With a promotion to sub-lieutenant he was next assigned as a fighter pilot with the 921st Sao Do (Red Star) Fighter Air Regiment of the Vietnamese Air Force, where he regularly flew as wing man for Vu Ngoc Dinh, one of the better-known fighter pilots.

Pham Tuan as a pilot with the 921st (Red Star) Fighter Regiment. (Photo: Vietnam Air Force)

During an air battle on 28 January 1970, Pham claimed to have shot down an F-4 Phantom jet, although this was not confirmed by U.S. records. Indeed the official records show that no Phantom jets were lost in combat between 20 January and 6 February that year. In 1971 he was promoted to full lieutenant and began operating on night sorties. In December 1972 it was reported that he flew back to his base, only to discover as he landed that the runway had been bombed and pitted with craters. His fighter jet hit a crater, flipped over, and skidded over 1,000 feet before finally rolling back over and coming to a halt. Amazingly, Tuan was not injured.

A week after his accident, on the evening of 27 December, Pham Tuan was ordered to take off from Yen Bai Airport in his MiG-21 No. 5121. "After taking off at night, I flew in the clouds at the altitude of 500 meters. The most experienced navigation officers … and those who were in charge of the flight map and air traffic directed me as if holding my hand, leading me to a good place to attack."[4] As he was flying south-west he saw and reportedly shot down a U.S. Air Force B-52 Stratofortress heavy bomber near Trung Quon, North Vietnam.

After launching two air-to-air missiles, Pham reported that he dived quickly to a safe altitude and landed. Several hours later, there was an official announcement that a pilot from the Vietnamese Air Force had shot down a B-52, and he received a congratulatory phone call from Vietnam's Minister of Defense. Thereafter, he was hailed as the first Vietnamese fighter pilot to have achieved this feat.

Pham's claim was later refuted by the U.S. Air Force, who stated that any B-52 heavy bombers lost over North Vietnam during Operation Linebacker II had been brought down by SA-2 surface-to-air missiles. The official records do show that B-52 Stratofortress 56-0674 *Ebony 2* from the 307th Strategic Wing, under the command of Capt. Frank Lewis, was "hit by SA-2: 2 killed, 4 POW. Claimed as kill by MiG-21 pilot". The SA-2 is a surface-to-air (SAM) missile, and Pham's claim was dismissed by the U.S. military and attributed to propaganda.

In recognition of his claim of downing the B-52 and for his heroism in defending Hanoi and Hai Phong, Pham Tuan was promoted to the rank of captain and received numerous decorations and awards, including the title of Hero of the Socialist Republic of Vietnam, the Order of Military Exploit 3rd Class, and the Order of Glorious Combatant 1st, 2nd and 3rd Class. He was further promoted to senior captain in 1974, and two years later he became deputy commander of the Red Star air regiment, with yet another promotion to major. He continued to serve with that regiment, eventually carrying out more than 200 combat flights in five years.

Around this time a pilot from another squadron invited Pham Tuan to come over and meet his sister, Tran Phuong Tan, a lieutenant in the Vietnamese People's Army Medical Corps. There was a mutual attraction and after a brief courtship they were married. In 1977, a year after he and his wife celebrated the birth of a baby daughter, Pham was sent to the Soviet Union for two years to attend the Yuriy Gagarin Air Force Academy. In 1979, still at the academy, he was promoted to lieutenant colonel and selected as a candidate for cosmonaut training in the Interkosmos program.

A SECOND RED STAR PILOT

Reflecting Pham Tuan's childhood, Bui Thanh Liem was also born into a poor family on 30 June 1949 in the Thuy Ai District of Hanoi City. His father, a member of the Communist Party of Vietnam who served as a company commander of the regimental intelligence in the Vietnamese Army, died fighting against French forces during the Indochina war (known in Vietnam as the Anti-French Resistance War) in 1950. His mother, Nguyen Ngoc Thi Yen (who had also joined the Communist Party), worked in the Thong Nhat Match factory in Hanoi.

Only one year old when his father was killed, Bui was raised by his mother. Details of his early childhood are scarce, but it is known that he graduated from high school with excellent grades in 1966. As the only son of a war hero he could have been exempted from military service, but with the encouragement of his mother he enlisted in the Vietnam People's Army in February 1966. He was subsequently assigned to the Vietnamese Air Force and joined the Ho Chi Minh Communist Youth Union in April 1967. He then traveled to Russia, where he received technical training at the Soviet College of Military Pilots, graduating with honors in November 1970. Back home again, he served as a sub-lieutenant pilot in the 921st (Red Star) Fighter Aviation Regiment, flying the highly-maneuverable MiG-21MF.

A portrait photo of Bui. (Photo: Vietnam Air Force)

On 27 June 1972, over Nghia Lo, Bui shot down an American F-4E belonging to the 366th Tactical Fighter Wing, flown by Maj. R. C. Miller along with his Weapons Systems Officer (WSO), 1st Lt. Richard McDow. Bui and Pham Phu Thai had maneuvered behind two of the Phantom jets. Pham shot down one with an infrared missile. Bui also fired an R-3S missile from 1,500 meters away and scored a direct hit. Both Americans managed to eject in time, and while Miller was subsequently rescued by a U.S. Search and Rescue team, McDow was captured before he could be extracted. As McDow later recorded:

> Our four F-4s were flying as a MiG CAP [Combat Air Patrol] for a SAR (Search and Rescue) mission after we had completed our primary mission of escorting two flights of F-4s into the Hanoi area for a chaff [radar fooling tinsel] drop. We were searching an area approximately 70 miles almost due west of Hanoi. During our search, two MiGs got behind us without our seeing them and fired at least two air-to-air missiles. The aircraft that I was in and our number four aircraft were hit. Our plane went out of control immediately and was tumbling and vibrating quite severely. There was no hope of regaining control so we (the pilot and I) ejected at approximately 5,000 feet above the ground. The crew of number four also ejected.[5]

Altogether, Bui was credited with shooting down two American aircraft, and in 1972 was promoted to senior lieutenant. In January the following year he joined the Communist Party of Vietnam. Promoted to captain in 1974, Bui was sent back to the Soviet Union to attend the Yuriy Gagarin Air Force Academy, graduating in 1978. Somewhere in this period he married Nguyen Thi Tuyet and they had a baby daughter, born in 1978. On his return to Vietnam he was promoted again to senior captain and appointed deputy chief of staff with the Red Star fighter-interceptor regiment.

THE FINAL TWO

On 17 May 1979, The Soviet Union and Vietnam signed an agreement of cooperation, under which Vietnam became the 9th Interkosmos nation to join the program. Vietnam's Ministry of Defense was then tasked with selecting four candidates from the Vietnamese Army, two of whom would be sent to train in Star City as cosmonaut-researchers for the Soyuz-37/Salyut-6 mission the following year.

At that point in history, Vietnam's leaders were desperate for some positive news to give a lift to their embattled population. Almost 30 years of ongoing combat against the French and Americans, amid a civil war, had been followed by brief but bloody battles with Cambodia and China. A huge push to collectivize farming was spreading hunger and destitution rather than Marxist ideals. That Vietnam should provide the first Asian to be launched into space as part of the Soviet Union's Interkosmos program, was seen by both nations as something that could provide a much-needed propaganda coup.

Very little is known concerning the Vietnamese selection process, but under strict Soviet qualification criteria the candidate pool was remarkably small, and only three serving pilots were found to be suitable. Vietnam's Ministry of Defense decided to expand the search by choosing a pilot from those studying in Russia at the Gagarin Air Force Academy. Pham Tuan had been nominated earlier, but rejected during the health screening process owing to the earlier issue of his heart problem. That had been hard for him to accept. However, in a twist of fate, with no suitable person located in Vietnam, his nomination was reconsidered and by virtue of his exceptional war record he became the fourth candidate.

The third but unsuccessful candidate was the highest-scoring ace of the Vietnam War, Nguyen Van Coc. (Photo: Vietnam Air Force)

Only one of the other two Vietnamese candidates has ever been identified. Col. Nguyen Van Coc was born in 1942. He joined the 921st Fighter Regiment in June 1965 and began operational flying in December 1965. With nine enemy planes and two drones confirmed he became recognized as the top ace of both sides in the Vietnam War. At the early stage of the Interkosmos selection process he was considered to be the principal candidate, but he failed to pass the final stage of the medical examination. He was later promoted to Commander of the Air Force and then Chief Inspector of the Ministry of Defense. He retired in 2002 after suffering from declining health.

Bui Thanh Liem (left) and Pham Tuan. (Photos: Spacefacts.de)

CONCENTRATED TRAINING

Initially, the two Vietnamese candidates were put through a familiarization course, studying and working in the Soyuz and Salyut space vehicle trainers and undertaking lessons to further their basic knowledge of the Russian language. Eventually they were paired off with two potential mission commanders to continue their training in parallel as two potential crews.

The two Vietnamese candidates in Star City pose with several cosmonauts and the Commander-in-Chief of the Vietnamese People's Army, Gen. Giap Vo Nguyen. From left: Pyotr Klimuk, Valeriy Bykovskiy, Georgiy Beregovoy, Bui Thanh Liem, Gen. Giap, Pham Tuan, Vladimir Shatalov and Aleksey Gubarev. (Photo: Author's collection)

Bui Thanh Liem was matched with three-flight veteran cosmonaut Valeriy Bykovskiy, who had gained international fame in 1963 by flying a five-day mission aboard Vostok-5, launching one day ahead of Vostok-6 with the world's first woman spacefarer, Valentina Tereshkova. Bykovskiy was a lucky man in April 1967. He was to have commanded the Soyuz-2 mission, but this launch was canceled after problems developed with Soyuz-1 in orbit. Later, the parachute system of that capsule failed and resulted in the death of the sole occupant, Vladimir Komarov. The investigation discovered the same fault in the Soyuz-2 spacecraft. If that mission had gone ahead, Bykovskiy and his crew would almost certainly have been killed on impact with the ground after re-entry. Bykovskiy made the eight-day Soyuz-22 Earth observation flight in 1976 along with flight engineer Vladimir Aksyonov, and two years later he commanded the Soyuz-31 Interkosmos mission which carried GDR cosmonaut-researcher Sigmund Jähn to a link-up with Salyut-6.

Pham Tuan and Soyuz-37 mission commander Viktor Gorbatko. (Photo: Author's collection)

Valeriy Bykovskiy and Bui Thanh Liem. (Photo: Spacefacts.de)

Pham Tuan was teamed with 44-year-old Viktor Gorbatko. On his first flight into space, Gorbatko was the research engineer aboard the three-man Soyuz-7 mission in 1969. Next he commanded the 18-day Soyuz-24 mission to the Salyut-5 'military' station in 1977 together with flight engineer Yuriy Glazkov.

The training as two-man crews began with splashdown and survival exercises on the Black Sea. In this, they were placed inside a mock-up of the Soyuz module which was lowered into the water from a ship. As the capsule bobbed up and down on the surface, small boats circled it, ready to help if the occupants got into any difficulty. Inside the capsule the crew members shed their space suits; a difficult task in the confined space, combined with the rolling motion of the capsule. Then they had to open the hatch in the neck of the capsule and jump into the water. After inflating small flotation devices, they would swim clear and ignite signal flares. A training session lasted several days, with tasks scheduled from early morning into the late afternoon, and such exercises continued through to mid-1980.

Gorbatko and Pham during Soyuz training. (Photo: Author's collection)

Pham Tuan had traveled to Moscow with his wife Tan and their three-year old daughter. In a recent interview he said his wife had found it hard at first to get used to life in Star City, but the wives of other cosmonauts helped her in her daily life while he was fully occupied in his mission training. After the two Vietnamese trainees had been paired with their respective Soviet commanders, one woman who became a close friend of Pham's family was Valentina Gorbatko, who entertained Pham's wife and daughter on weekends with traditional Russian dishes including salad, pickled cucumbers, and goose with apples. She also taught Tan how to make pasta and bake cakes and they would often go on sightseeing trips around Moscow.

When the prime crew was selected, it was Gorbatko and Pham Tuan.

AN OLYMPIC PROPAGANDA MISSION

The Soyuz-37 mission began with a trouble-free lift-off of the Soyuz-U carrier rocket from the Baykonur launch pad at 18:33:03 (UTC) on 23 July 1980 – the same launch pad used to send Yuriy Gagarin on his historic flight in 1961. Looking back in July 2000, Pham Tuan recalled what he said was "the greatest day of my life. As everybody watched us enter the spacecraft at Baykonur Cosmodrome, I felt almost as if time was standing still. I grew more and more anxious while listening to the countdown, even though I had been training for that moment."

Predictably, before the lift-off, Pham Tuan had said, "We have enjoyed the kindness and warmth of the Soviet people ever since our first minutes in Star City. They welcomed us extremely cordially. Space exploration is a totally new field for us, and our Soviet comrades gave us tremendous assistance in sorting out all the complexities."

As usual, details of the launch were not released by the Soviet news agency TASS until the spacecraft was safely in orbit. Controllers then felt sufficiently confident to announce via TASS that veteran Soviet pilot Viktor Gorbatko and Lt. Col. Pham Tuan would dock with Salyut-6 and work in orbit with the resident crew of Leonid Popov and Valeriy Ryumin.

Word of the launch was circulated among East European correspondents in Moscow well before it was announced by TASS, and apparently well before the actual launch, indicating confidence that all would go well. TASS reported that the latest international flight was "a vivid example of the fraternal friendship and close cooperation between the peoples" of Vietnam and the Soviet Union.

In a statement released by TASS after the launch, Pham spoke of the strong symbolism involved for Hanoi. "This flight is being carried out in an anniversary year for my country," he said, "when the entire population is celebrating the 50th anniversary of the founding of the Communist Party of Vietnam, the 90th anniversary of the beloved leader Ho Chi Minh, and the 35th anniversary of the proclamation of Vietnam's independence."

Although another international flight had been expected, the selection of a Vietnamese cosmonaut-researcher surprised many Western space analysts who closely watched Soviet space activities. They knew the names of the Bloc countries which had pilots undergoing mission training, and fully expected the next Interkosmos spacefarer to be from Romania, Cuba or Mongolia. They also knew that Vietnam was very much the new kid on the block insofar as Interkosmos was concerned.

There was, however, a reason why the Vietnamese mission basically jumped ahead of those of the other Bloc nations. In the summer of 1980 the Olympic Games were being held in Moscow. Normally a time for celebration of athleticism and brilliance in various sporting disciplines, these Games were badly tainted by the Soviet invasion of Afghanistan the year before, and this had prompted several nations to either boycott the Games or to advise their athletes not to attend in protest. U.S. President Jimmy Carter even said that if Soviet troops were not withdrawn in a month he would ask the United States Olympic Committee to urge the International Olympic Committee to transfer or cancel the Games in Moscow. When the event went ahead, it was with a number of national and individual athlete boycotts. Coming just five years after the reunification of Vietnam and the end of the long and bitter war with the United States, the Soviet-Vietnam mission involving Pham

Tuan visiting Salyut-6 would be seen as a great propaganda victory for the communist government, not only spotlighting the troubled Olympics but also dramatizing Soviet ties to Third World nations.

Writing on these Olympic Games, analyst Barukh Hazan stated, "Soviet space research featured largely in the Soviet propaganda effort during the Games. During the opening ceremony the huge film screen of Lenin Stadium repeatedly showed the faces of Leonid Popov and Valeriy Ryumin, the Salyut-6 crew, as well as their greeting to participants in the Olympic Games: 'Let the Olympic flame of friendship always burn, let rivalry be confined to the sports field only.' Soviet cosmonauts visited the Olympic village and foreign tourists were brought to the space exhibit. The climax was the launching of the spaceship with its joint Soviet-Vietnamese crew on 24 July.

"The event was widely reported and commented on by the Soviet and East European mass media, its relevance to the Games being explained in the following manner: 'Only a country with a great industrial, scientific, technological, intellectual, and moral potential could organize such unrepeatable Olympic Games. Many people asked themselves: How is it possible simultaneously to organize such huge Games of a Soviet scale and scope, and launch such a complex cosmic flight with an international crew in which there is a representative of the Yellow Continent … This became an exceptional example for the thousands of foreign journalists and tourists, who could learn the truth about the USSR.' A French journalist was quoted as adding: 'The international Soyuz-37 crew is an example of Soviet assistance to developing countries.'"[6] The mention of the Yellow Continent here referred to the yellow ring of the Olympic flag, which represented Asia.

WORKING IN ORBIT

During their first three orbits around the Earth, Pham and Gorbatko checked the equipment and systems of their spacecraft to ensure that all was in readiness for the docking procedure. After receiving permission from the ground, they shed their cumbersome Sokol suits and were able to relax in lightweight clothing. On the 16th and 17th orbits Gorbatko began the automatic sequence that would allow the docking to take place.

Soyuz-37 docked at the aft port of the Salyut-6/Soyuz-36 complex at 23:02 Moscow time on 24 July 1980. This was preceded on 18 July by the separation of the automatic cargo craft Progress-10, which had been docked at that port since 29 June.

With a successful link-up completed, Gorbatko and Pham made preparations to spend the next seven days working in orbit around the Earth with the station's occupants Leonid Popov and Valeriy Ryumin. On entering the space station they received the traditional greeting of bread and salt.

Popov and Ryumin had been aboard Salyut-6 since 10 April and were hoping to break the 175-day record set by Ryumin and Vladimir Lyakhov the previous year. It was not originally intended that Ryumin should undertake two successive long-duration missions, but he had volunteered when Valentin Lebedev was injured during training.

Despite the very obvious propaganda benefits derived from this international flight, some useful experimentation and Earth observation was carried out by the Soyuz-37 crew.

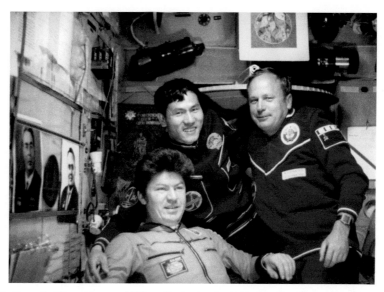

Valeriy Ryumin, Pham Tuan and Viktor Gorbatko aboard Salyut-6. (Photo: Spacefacts.de)

One of their main assignments was to study the effects of chemical warfare on nature and agriculture in Vietnam during the 1960s. In this experiment, called Halong, Pham and Gorbatko studied various areas of nature across the country, while the Mekong River delta and the Tai Nguen Plateau were scrutinized using the MKF-6M multi-spectral camera and the Bulgarian Spektr-15 spectrophotometer. The images and data would later assist in determining the feasibility of reclaiming forests and rice paddies in areas that were badly damaged by the extensive use of chemical weapons. The data also provided the basis for compiling a true geographic map of Vietnam and enabled reserves of coal, oil and natural gas to be sought.

During the first days aboard the station, in his free time Pham would sit by the window with a pair of binoculars as they passed over Vietnam, not wanting to leave it until he saw his beloved village. He often ignored the timetable by which he should be sleeping, so he could sit at the command center and silently watch the Earth pass below.

Meanwhile the crew also conducted experiments using the floating azolla plant, which is extremely rich in nitrogen and normally multiplies rapidly in paddy fields in order to supply them with much-needed nitrogen. The experiment would test the plant's ability to grow in weightlessness, as a potential food source on long-duration space missions.

Engineering studies were also carried out. These had been set up by Dr. Tran Huam Hoai of the State Research Center of Vietnam, working in collaboration with scientific collectives from the German Democratic Republic. In one, samples were manufactured and packed in special siliceous ampoules that were melted in the Kristall furnace so that the earlier created crystal nucleus was not damaged. The ampoules were cooled by slowly withdrawing them from the chamber. In this manner, the nucleus could be transformed into crystal alloys of a predetermined shape.

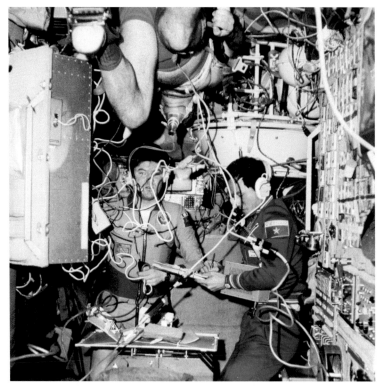

Amid a tangle of wires, Ryumin and Pham observe an upside-down Gorbatko during a medical monitoring experiment. (Photo: Author's collection)

Medical experiments were carried out, as they were on all of the Interkosmos flights. The crew had to record answers to questions relating to their mental health, hearing, respiration, oxygen absorption in weightlessness, discerning different tastes using an electrogustometer, and report on their work capacity.

During his eight-day sojourn aboard Salyut-6, Pham beamed back messages hailing his country's long struggle for independence, and thanking the Communist party "for having trained me and given me wings to fly into space". But his hungry compatriots weren't so easily taken in. A popular rhyme at the time pondered, "We have no rice, we have no noodles, so why are you going into space, Mr. Tuan?"

HOMEWARD BOUND

The plan was for Gorbatko and Pham to return in Soyuz-36, leaving the fresher Soyuz-37 vehicle for Popov and Ryumin. This change-over entailed swapping couches and personal effects from one spacecraft to the other. When this was done, the scientific material which had been gathered was carefully loaded into the Soyuz-36 descent module.

On 31 July, Soyuz-36 undocked from Salyut-6 and landed the same day 112 miles south-east of Dzhezkazgan after a mission lasting 188 hours and 42 minutes. A medical check-up carried out immediately after the flight indicated that both men were in excellent condition.

Safely back on Earth, Pham and Gorbatko relax after their space journey. (Photo: Author's collection)

Pham Tuan often recalls arriving back on terra firma. "I still remember my first landing on Earth after going to space," he said in a 2014 interview. "I felt like screaming 'Hooray! We're alive! We succeeded!' Then our capsule bounced several times before it came to a halt."[7]

The day after Gorbatko and Pham Tuan's return to Earth, Popov and Ryumin undocked Soyuz-37 and maneuvered it to the forward end of the station for the rest of their mission.

When Gorbatko and Pham Tuan returned to Star City, there was a moving meeting with Pham's father. And then there were the medal presentations at the Kremlin. "In the grand ballroom and with the participation of some top leaders and diplomats, General Secretary of the Communist Party of the Soviet Union Leonid Brezhnev presented the Order of Lenin to Gorbatko," Pham Tuan recalled. "After that, he read the decree for my award. Mr. Brezhnev pinned a Gold Star for the Hero of the Soviet Union and the Order of Lenin on my chest. He then kissed and hugged me like he was my father."

Several days later, Pham Tuan, Viktor Gorbatko and their families flew to Vietnam, taking with them new records and pictures of that country taken from orbit. The road from Gia Lam airport to Hanoi, five miles long, was crowded with people, banners, and flowers. Words of greeting and congratulation could be heard everywhere. They were welcomed at the gate of the Presidential Palace by General Secretary of the Communist Party of

The four Soyuz-37 crewmembers at an official function in Hanoi. (Photo: Author's collection)

Vietnam Le Duan, State President Nguyen Huu Tho, Prime Minister Pham Van Dong, and Minister of Defense Van Tien Dung. It was 2 September, their national day, 35 years after Vietnam had declared independence. The National Assembly of Vietnam conferred the title of Hero of Socialist Vietnam and Gold Star on Viktor Gorbatko, who already held the title of Hero of the Soviet Union (bestowed twice) and on Pham Tuan, who now had two Gold Stars. The awards, the citation said, "were for their exemplary performance of tasks aboard the complex [in space] and for courage and heroism shown during the flight".

Thirty years on from their space mission, Pham is reunited with Viktor Gorbatko. (Photo: Author's collection)

For several days after that, the two crewmen attended meetings in cities and villages all over Vietnam, and were welcomed as heroes. Pham Tuan invited Gorbatko's family to his hometown where, as a boy from Quoc Tuan village, Pham used to gaze into the sky, which was daily crossed by American planes, and hope to one day become a pilot. The two men also planted trees as keepsakes in several places they visited. Years later, when Gorbatko once again visited Vietnam, he sat in the shade of one of the trees he had planted.

Following the successful flight of Soyuz-37, Pham's back-up pilot, Bui Thanh Liem, returned to Vietnam and resumed his career in the nation's air force, now promoted to the rank of major. He was awarded the Medal of Ho Chi Minh City for having successfully completed training for the Soviet-Vietnamese space flight. A year later, on 26 September 1981, he was operating a MiG-21 during a training flight over the Gulf of Tonkin off the coast of North Vietnam when his fighter jet crashed into the sea. It was never recovered. Just 32 years old, he left behind his wife and a three-year-old daughter.

Pham Tuan with his wife Tan. Behind him is a photo of Pham interviewing a downed American pilot during the Vietnam War. (Photo: Vietnam News)

Pham Tuan retired in early 2008 having attained the rank of lieutenant general as Head of the General Department of Defense Industry of the Ministry of Defense. At the time he said, "Now I've really landed. My motto is whatever we do, we should love it and pursue it to the end of the path." His wife Tan, who once served in the battlefields, still works as a military doctor. Their daughter Tkhu, born 1976, completed her post-graduate training in auditing and received her Master of Arts degree in Australia. Their son, Pham Tuan Anh, is now in the process of completing a post-graduate course in the United Kingdom.

REFERENCES

1. *Vietnam News* newspaper, Hanoi, "Vietnam Celebrates Foray into Space," issue 24 July 2000
2. Global security.org undated article, "Vietnamese Cosmonauts," available from website: *http://www.globalsecurity.org/space/world/vietnam/piloted.htm*
3. Gordon R. Hooper, *The Soviet Cosmonaut Team, Volume 2: Cosmonaut Biographies* (revised edition), GRH Publications, Suffolk England, 1990, pg. 309
4. Vietnamese Ministry of Defense, *People's Army Newspaper*, article, "Shoots down B-52, gets 'ticket' to space," issue 8 May 2008
5. Richard A. McDow, from *We Came Home*, P.O.W. Publications, Toluca Lake, CA, 1977
6. Barukh Hazan, *Olympic Sports and Propaganda Games: Moscow 1980*, Transaction Publishers, New Brunswick, NJ, 1982
7. VIETMAZ: Vietnam Local News, "Vietnam and Sweden's first astronauts say dreams must go with effort," Tuoitrenews, 22 November 2014

8

The Cuban Salyut mission

Arnaldo Tamayo Méndez, a 38-year-old Cuban-born pilot of African descent, became the first Latin American and first black person to fly into space when he was launched on the Soyuz-38 Soviet-Cuban mission in September 1980. In speaking of his life and the many opportunities it presented, Méndez later pointed out he owed everything to the revolution which brought Fidel Castro to power in Cuba in 1959.

"I had dreamed of flying since I was a child," he mused. "But before the revolution all paths into the sky were barred because I was a boy who came from a poor black family. I had no chance of getting an education."[1]

A CHILD OF CUBA

Arnaldo Tamayo Méndez was born into a poverty stricken family on 29 January 1942 in Baracoa, Guantánamo, near the remote eastern tip of Cuba. Before he had even reached his first birthday he was orphaned when both of his parents died due to an undisclosed "serious illness." He was subsequently adopted by his maternal uncle Rafael Tamayo and his wife Esperanza Méndez, who also lived in poverty. According to Spanish naming customs, the boy's first or paternal family name was Tamayo, and his second or maternal family name was Méndez, hence Arnaldo Tamayo Méndez.

In 1952, when Arnaldo was just 10 years old, a military officer named Fugencio Batista orchestrated a coup that removed the President of Cuba, Carlos Prio, from office. Elections were abandoned. The future looked dismal for a youth coming from a poor family entering his second decade of life in a society being run in the interest of the wealthy by a military dictator. He may have been too young to be aware of the challenge to dictator Batista when, on 26 July 1953, a passionate young Cuban lawyer named Fidel Alejandro Castro Ruz and about 140 rebels attacked the federal garrison in Moncada. Even though this operation was well-planned and had the element of surprise, it was a near-total failure owing to the greater numbers and weapons of the army soldiers and some remarkably bad luck on the part of the attackers, many of whom were captured, tortured and executed. Fidel Castro and his brother Raúl were put on trial. They may have lost the battle but eventually they won the war – the assault on Moncada was the first armed action of the Cuban Revolution which triumphed in 1959. Arnaldo Tamayo Méndez was growing up in turbulent times.

© Springer International Publishing Switzerland 2016
C. Burgess, B. Vis, *Interkosmos*, Springer Praxis Books, DOI 10.1007/978-3-319-24163-0_8

"I was born in a poor negro family," he reflected, "and although I dreamed of flying from childhood, that was utterly unthinkable. I had to start work early, as a street shoe-shine boy – a poor man's profession – or selling vegetables." After finding employment delivering milk for a dairy and as a low-paid laborer in a furniture factory, by the age of thirteen he was an apprentice carpenter. "The situation completely changed after the revolution." During the Cuban Revolution in the late 1950s he joined the Association of Young Rebels, which had protested against the Batista regime, and was working in the Sierra Maestra mountains. In December 1960 he began attending the Rebeldi Technical Institute, where he took a course for aviation technicians. After graduating the following year with top marks he resolved to become a pilot. "Also dreams of the cosmos, but I understand that is impossible, for small Cuba that is not possible, the conquest of space."[2]

First Cuban cosmonaut: Arnaldo Tamayo Méndez. (Photo: Author's collection)

BECOMING A PILOT

After graduating from the Rebeldi Institute, Tamayo Méndez had been selected to undertake further studies in the Soviet Union in order to fulfill his ambition of becoming a pilot. Once there, he was initially restricted to being an aircraft technician owing to unspecified medical reasons. But it wasn't long before the Soviet doctors gave him permission to undergo flight training. From April 1961 to May 1962 he undertook an intensive one-year course of study at Yeisk Higher Military Aviation School, where he trained to fly the MiG-15 fighter jet. On hearing of the orbital flight of cosmonaut Yuriy Gagarin on 12 April 1961, he reaffirmed his determination to make aviation his career.

Returning to Cuba in 1962, Tamayo Méndez joined the Playa Girón Brigade of the Cuban Revolutionary Guard as a flight instructor. That October saw Cuba and the Soviet Union on one side pitted against the United States in a crisis that took the world perilously close to the brink of a thermonuclear war. He is said to have played an important role during this Cuban Missile Crisis, flying twenty reconnaissance missions as a squadron leader and deputy wing commander, and repeatedly intercepting American aircraft that had violated Cuban airspace.

In December 1967, Tamayo Méndez married Maria Lobaina, and they would later have two sons, Orlando and Arnaldo. That same year he became a member of the Communist Party of Cuba and was sent to Vietnam to study that nation's conflict with the United States, serving with the Cuban Revolutionary Armed Forces. On his return to Cuba in 1968 he was promoted to military pilot 1st class.

From 1969 to 1971 he studied at the Maximo Gomez Basic College of the Revolutionary Armed Forces, receiving advanced command and staff courses, and from 1971 to 1975 was chief of staff of a fighter brigade while continuing to fly as a pilot. He then became alternate chief of staff to the Aviation Brigade of Santa Clara, and the following year was promoted to the rank of captain.

THE CALL OF THE COSMOS

According to Torres Yribar, President of the Cuban Academy of Sciences, Cuba began its space research activities in 1964, and became a member of the Interkosmos program three years later. Scientists in Cuba had already been analyzing data provided by meteorological satellites to assist them in monitoring industrial pollution and the effects of climate on sugar cane production.

The selection process to identify the first Cuban cosmonaut had begun in 1976. Initially, the records of 600 armed forces personnel were examined, resulting in 80 being set aside for further examination as potential candidates. Following several rigorous medical and physical performance tests at the Instituto Superior de Medicina Militar (named after Dr. Luis Diaz Soto) in Havana, this number was reduced to 20, then to nine by the end of 1977, and finally to four. These men then went to the Soviet Union for a final and specialized appraisal over a four-week period.

Following this appraisal, the joint Soviet-Cuban commission chose two candidates, both military pilots: Arnaldo Tamayo Méndez, who was then a staff officer with a fighter brigade, and José Armando López Falcón, a captain in the Cuban Air Force. They were notified of their selection on 1 March 1978, with one destined to become the seventh cosmonaut of the Interkosmos program. As Tamayo Méndez recalled, "Our preparation officially began on March 23, 1978."

The other Cuban candidate, Capt. José Armando López Falcón, was born in Havana, Cuba on 8 February 1950. His father worked for the Ministry of the Food Industry and his mother for the Ministry of Transport. In 1966, at the age of 16, he entered the Carlos Ulloa

José López Falcón (Photo: Author's collection)

Military Jet Fighter School. A year later, on 5 July 1967, upon his graduation from fighter school and commissioning as a junior lieutenant, he was assigned as a flight instructor. He was also a member of the Union of Young Communists. The following year he was sent to the Soviet Union to study at a military flight school, where he learned to fly modern fighter jets, and upon his return he was named commander of an air regiment unit.

In 1971 López Falcón became a lieutenant and in 1973 senior lieutenant. Three years on, he was sent to the Soviet Union once again to further improve his skills. After returning, he served in a major air force unit. While flying in a MiG-21 as a pilot 1st class he was several times involved in intercepting aircraft that had violated Cuban airspace. In 1978, and now a captain, he was selected as a candidate cosmonaut-researcher. In April the following year he joined the Cuban Communist Party. Like his mother, his wife Falcón Kintero Iraida worked at the Ministry of Transport. Their daughter Daira was born in 1974. While preparing for the Soyuz space mission, López Falcón continued to study at the Electronic-Mechanical Faculty of the Jose Antonio Echeverria Polytechnic University in Havana.

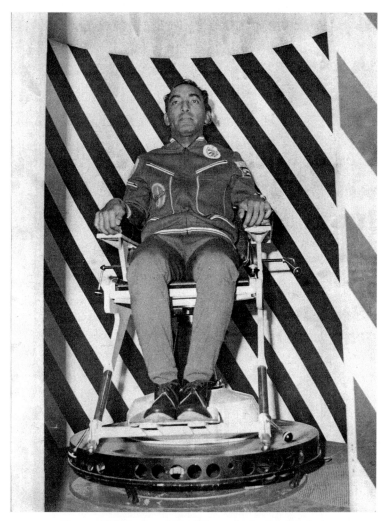

Tamayo Méndez in training. (Photo: Author's collection)

After several months of training, the Interkosmos candidates were paired off with Soviet commanders. Tamayo Méndez got Col. Yuriy Romanenko who had previously flown on the Soyuz-26 mission to Salyut-6 in December 1977 and, along with engineer Georgiy Grechko, had spent 96 days aboard the station, during which they conducted a 90-minute spacewalk.

López Falcón was paired with another veteran. Col. Yevgeniy Khrunov was the research engineer on the three-man Soyuz-5 crew that was commanded by Boris Volynov and docked with Soyuz-4 on 16 January 1969. Along with flight engineer Aleksey Yeliseyev, Khrunov had spacewalked to Soyuz-4 and joined Vladimir Shatalov for the journey home.

The two Soyuz-38 crews: Yuriy Romanenko and Arnaldo Tamayo Méndez (front) and Yevgeniy Khrunov and José Armando López Falcón. (Photo: Author's collection)

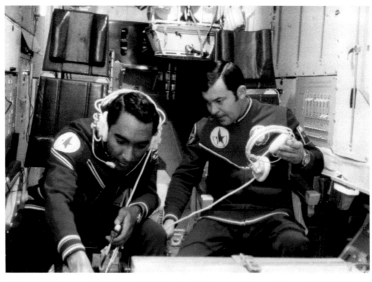

Romanenko and Tamayo Méndez train aboard a Salyut mock-up. (Photo: Author's collection)

Khrunov and López Falcón in a Soyuz mock-up. (Photo: Author's collection)

Water survival training on the Black Sea. (Photo: Author's collection)

"What I like about Arnaldo Tamayo is his exceptional sense of purpose," observed Yuriy Romanenko of the Cuban pilot who became his Soyuz partner. "He is a very determined individual, but also very kind and has a sharp sense of humor. Of course, as a real Cuban, he is temperamental but is in good command of his emotions. In training Arnaldo operated with a calculating and cool mind. We have done some hard work together, spending about 1,500 hours in all in the classroom and in practical training."[3]

Soyuz-38 prime crew: Tamayo Méndez and Romanenko. (Photo: Author's collection)

It is believed that Fidel Castro wanted Méndez to fly the Soviet-Cuban mission and used his considerable influence to ensure that this was the outcome. The Romanian Interkosmos candidate Dumitru-Dorin Prunariu has said that he had trained at the same time as Tamayo Méndez and while the choice was supposedly left up to the Cuban authorities, he felt there were far more political issues associated with the eventual selection. "Both candidates were very well trained, very well prepared, but Tamayo was the first black person to be sent into space. That was a political challenge for Russia … for the Soviet Union in that time, to send him. He was well prepared, no doubt, but his Russian was pretty poor, compared to the other one [José López Falcón]."[4]

The back-up crew of López Falcón and Khrunov. (Photo: Spacefacts.de)

The rocket that will lift Soyuz-38 into orbit heads for the Baykonur launch pad. (Photo: Author's collection)

JOINT SOVIET-CUBAN MISSION

Tamayo Méndez and Soviet cosmonaut Yuriy Romanenko flew together aboard Soyuz-38, which lifted off from Baykonur on 18 September 1980. Two days later they docked at the Salyut-6 space station.

Cuban President Fidel Castro watches the mission report on television. (Photo: Author's collection)

It was Tamayo Méndez who first floated through the hatches to greet Leonid Popov and Valeriy Ryumin. After the traditional welcoming ceremony of sharing bread and salt, they received congratulatory messages from Soviet President Leonid Brezhnev and from Cuban President Fidel Castro, both of whom stressed the point that the joint space venture would enhance mutual goodwill between their nations. For the television audience back on Earth, and in no obvious hurry to commence their serious scientific program, the newcomers then proceeded with a series of somersaults to demonstrate their enjoyment of the experience of weightlessness.

While aboard the station, Romanenko and Tamayo Méndez carried out 27 experiments prepared jointly by Soviet and Cuban specialists. The flight pattern of the mission was in part prepared by the National Academy of Sciences of Cuba. It included materials science experiments such as growing layers of gallium arsenide in a furnace and the cultivation of some organic monocrystals, grown for the first time in weightlessness, using Cuban sugar syrup. The medical experiments studied changes in the structure and function of the human foot arch in weightlessness, the human central nervous system, psychomotor coordination between the left and right hands, and space adaptation syndrome (widely known as 'space sickness'). There was also an intra-cellular investigation of rapid-growth yeasts. In addition, they carried out studies from space of the territory of the Caribbean island and its associated continental shelf in search of minerals and possible oil deposits, and continued experiments initiated by previous international crews.

The four cosmonauts aboard Salyut-6. (Photo: Author's collection)

At one stage, Tamayo Méndez peered down at his Cuban homeland, remarking that its shape reminded him of "a large fish with fins". During one cloudless night he was able to observe the shimmering city lights of Havana.

The four cosmonauts spent Saturday, 20 September relaxing aboard on a scheduled rest day. At one stage Tamayo Méndez complained that he didn't feel as well as he had felt on the first two days of the mission, and physicians advised him to curtail strenuous activities.

Another joint Soviet-Cuban message was sent to the visiting crew to congratulate the two men and stress the importance of the Interkosmos program in the exploitation of outer space for peaceful purposes. "The results of your work in space will make a big contribution to the progress of science and technology, [and] will facilitate the fulfillment of economic tasks in the USSR and Cuba and in all the fraternal socialist countries."[5]

On 26 September the Soyuz-38 spacecraft touched down around 110 miles south-east of Dzhezkazgan, missing its planned landing target by a mere two miles. In total, the crew had spent 7 days, 20 hours, 43 minutes and 24 seconds in space. Because it landed shortly before 7:00 p.m. Moscow time, it was necessary for the recovery crews to use searchlights to locate the capsule.

The Soviet-Cuban mission to Salyut-6 had social, political, and economic value beyond the pursuit of scientific knowledge. Soviet President Leonid Brezhnev and Cuban President Fidel Castro endorsed the joint mission saying that it strengthened the bond between the two nations.

Yuriy Romanenko and Arnaldo Tamayo Méndez were both awarded the title of Hero of the Soviet Union and the Order of Lenin. In addition, Tamayo Méndez received the title of the Hero of the Republic of Cuba, with Gold Star Medal, the Order of the Bay of Pigs, and the Order of Playa Girón. His triumph was celebrated by the issuing of a stamp bearing his picture.

On the 30th anniversary of his space flight, Arnaldo Tamayo Méndez examines the Soyuz-38 descent module in the Room of Cosmonautics in the Guantánamo Museum of the Revolution. (Photo: RadioRebelde)

These days his space suit is preserved at the Museum of the Revolution in Havana, while some of the equipment he used on the flight, as well as the descent capsule of the Soyuz-38 spacecraft, are on display in the Room of Cosmonautics in the Provincial Space Museum in his native Guantánamo.

The prime crew members of the first seven Interkosmos missions during a trip to the Kosmos Pavilion at the VDNKh (All-Russia Exhibition Center) in Moscow, 1981. From left: Tamayo Mendez, Sıgmund Jähn, Vladimir Remek, Miroslaw Hermaszewski, Georgi Ivanov, Bertalan Farkas and Pham Tuan. (Photo: Author's collection)

LIFE AFTER SPACE

Tamayo Méndez later served as vice-president of the Society of Cuban-Soviet Friendship. In 1980 he became a member of the National Assembly of People's Power, the elected Cuban national legislature, representing his home municipality of Baracoa. From 1981 to 1992 he directed both the Society of Military and Patriotic Education (SEPMI) and the Aviation Club of Cuba. During that time he was promoted to the rank of brigadier general in the Cuban Air Force. He has also served as director of Cuba's civil defense organization.

President Fidel Castro took great political advantage from the Soviet-Cuban flight, and in a passionate, rhetoric-laden speech at the Conventions Palace on the outskirts of Havana, he not only emphasized Cuba's affection for the Soviet Union and its people, but took political advantage of Tamayo Méndez's African ancestry.

> For our people, the cosmonauts represent the Soviet man, the best fruits of the Soviet revolution. If Lenin's generation carried out the revolution and the next generation defended the fatherland and defeated fascism, this generation is characterized by the great advances and great scientific and technical gains and for having developed the daring men who have conquered space. When Gagarin predicted that not many years would go by before Cuban cosmonauts traveled in space, who would have imagined that nineteen years later, on a day like today, we would meet to commemorate, honor, and pay tribute to the Soviet-Cuban crew which has made it possible for us to say that our country has already sent a man into space.
>
> A cosmonaut is not chosen by chance. Tamayo said here that he feels honored, very honored, because our party and government chose him to be the first Cuban cosmonaut. That is not the way it happened. I repeat that a cosmonaut cannot be chosen by chance. Exceptional conditions are required for this mission. A great temperament is required. Great talent is required. Great courage, great coolness is required. A revolutionary attitude is required. Very high morals are required. It is required to be an example. In a few words, it is necessary to be a communist.
>
> I recently said, and now wish to reaffirm, that revolutionary virtues, courage and many of the conditions represented by Comrade Tamayo are precisely the virtues of our people. I said that there could be millions of Tamayos in our country, that I was sure of that because I really am sure of that. He is a symbol of the temperament, determination, audacity, courage, talent, and revolutionary spirit of our people.
>
> He symbolizes our heroic combatants, the heroic combatants who gave their lives for the revolution's triumph, for the defense of the revolution. He symbolizes the heroic internationalist combatants of our people. He symbolizes our heroic internationalist workers. He symbolizes the vanguard members of our working class. He symbolizes exemplary workers. He symbolizes the work heroes …
>
> With the revolution the doors opened up for him as they did for all our youth, as they did for all our people. The opportunity to study, the opportunity to excel, the opportunity to serve his people were some of his options as a humble youth. And his humble beginnings has been referred to, repeated and insisted on because it really constitutes a symbol, the fact that our first cosmonaut and the first cosmonaut from Latin America is the first cosmonaut from Africa. It is not our whim to say that he also is the first cosmonaut from Africa because Tamayo, an eminently black man who also has in his veins Indian and Spanish blood is a symbol of the blood which,

as demonstrated by the most severe test of our fatherland's history, gave rise to our people. It is African blood. It is Indian blood. It is Spanish blood.

That is why we say he also symbolizes Africa because he is the first descendant of Africa to travel in space. It is a symbol that a man from such humble origin has attained such extraordinary success. Of course, only the revolution – and the revolution alone – has made it possible for a youth such as Tamayo to have that possibility.[6]

In 2009, Tamayo Méndez was appointed head of the Department of Foreign Relations of the Ministry for the Cuban Revolutionary Armed Forces.

The Russian Order of Friendship, granted by Russian President Dmitriy Medvedev, was awarded to Tamayo Méndez in April 2011 by Ivan Melnikov, vice president of the Duma. Melnikov said the award "reflects the great contribution of Tamayo to the relations between our two nations, as the first cosmonaut of the island-nation and [also] as the president of the parliamentary group Cuba-Russia Friendship." In 2014 Tamayo Méndez released a Spanish-language book of his experiences as a cosmonaut titled, *Un Cubano en el Cosmos* (A Cuban in the Cosmos).

A recent photo of José López Falcón. (Photo: Unaccredited)

For his part, José López Falcón returned to Cuba following his role as back-up pilot and spacecraft communicator in the Flight Control Center for the Interkosmos flight. He then continued his service as a captain in the Cuban Air Force, with special emphasis on flight operations and security. Today he is serving as the Technical and Training Officer for the International Civil Aviation Organization (ICAO), with responsibility for their universal safety oversight program. He is married with one child, a daughter born circa 1974.

Yuriy Romanenko flew into orbit once more on the Soyuz TM-2 mission in 1987. This was the first manned use of this new Soyuz spacecraft, and he commanded the second crew to the Mir station. On retiring from active space flight in 1988 he was appointed director of the Buran program, which was intended to develop a Soviet rival to the U.S. space shuttle.

Following his back-up command duties for Soyuz-38, Yevgeniy Khrunov was offered command of another Interkosmos flight, this time teamed with a Romanian cosmonaut, but it is known that he lost this flight due to disciplinary reasons. Khrunov subsequently resigned from the cosmonaut team on Christmas Day 1980, and later became chief of administration for the USSR State Committee for Foreign-Economic Relations. He suffered a heart attack and died in Moscow on 19 May 2000, aged 67.

SOYUZ-38 MAJOR EXPERIMENTS SUMMARY

Cortex: This was designed to study the electrical response of the brain to certain stimuli. Scientists wanted to analyze how brain activity was modified in space and whether or not these changes were reversible after returning to Earth. A special helmet was made for each cosmonaut in which silver electrodes were installed, as well as a headset, sound and light stimulators, amplifiers to stimulate brain electrical activity, and a tape recorder.

Support: Changes in the structure and function of the foot were known to make walking a little different on the cosmonauts' return to Earth. For four hours per day, Cuban cosmonaut Tamayo Méndez donned special shoes named Dome-Sand-501 in an attempt to determine the cause of the changes in the movement of their feet after a period of prolonged weightlessness. These results would allow advances to be made in the design of more appropriate footwear to preserve the structure of the foot in weightless conditions.

Stress: This experiment prepared by the Cuban Institute of Endocrinology and Metabolic Diseases was aimed at recognizing whether the effects of psychological stress experienced during mission preparation, during launch, in weightlessness, and over many continuous working hours, caused hormonal changes in an individual. It was based on an analysis of blood and urine samples taken from the crew before departure and following their return.

Anthropometry: An experiment conducted to determine changes in the levels of skeletal muscle structure in cosmonauts. The tests were performed through a special instrument known as the Calibrómetro Cosmos-726, designed and built by Cuban specialists, which assessed the adipose tissue and the stimulation of body fat.

Blood Circulation: This experiment investigated the effects of weightlessness on the human circulatory system. Tests had been carried out on previous flights and were repeated during this mission using a Chibis suit – a below-the-waist reduced-pressure device. Crew members completed exercise protocols wearing the Chibis to provide gravity-simulating stress to the cardiovascular/circulatory system and re-establishing the body's orthostatic tolerance (where a loss of blood to the brain causes a person to faint) after extended periods of microgravity. Negative pressure on the legs causes blood to accumulate in the lower extremities, which is the case in a gravity environment. Orthostatic intolerance has been a frequent complaint in humans returning from long-duration space flights.

Hatuey: This experiment was an attempt to learn more about the effects of weightlessness on the process of cell division in the living organism. Yeast was sent up to Salyut-6 and then its process of cell reproduction was observed.

Immunity: Scientists analyzed blood samples before and after the flight to determine what happens to the human immune system and the concentration levels of antibodies and other proteins and minerals under conditions of weightlessness.

Balance: The loss of water, fat, and minerals in cosmonauts was studied during their stay in space. In this case the study became more interesting as Tamayo came from a tropical island country and savanna, indicating a very different situation to that in previous experiments with cosmonauts from the Soviet Union or other countries in that part of the world.

Sugar: This experiment studied the growth of a single crystal of sucrose in weightlessness.

Tropic III: For this experiment the MKF-6M multispectral camera was used to photograph selected areas around the island of Cuba. Because Salyut-6 traveled in an orbit which was steeply inclined to the equator and Cuba extends from east to west, Tamayo Méndez had to optimize his use of the time available.[7]

REFERENCES

1. Mikhail Chernyshov, article, "Cuba in Outer Space," *Soviet Weekly* newspaper, issue 2 October 1980
2. *Soviet Weekly* newspaper biography, "Arnaldo Tamayo Mendez," issue 9 October 1980
3. Mikhail Chernyshov, article, "Cuba in Outer Space," *Soviet Weekly* newspaper, issue 2 October 1980
4. Bert Vis interview with Dumitru Prunariu, 22nd Planetary Congress of the Association of Space Explorers, Prague, Czech Republic, 6 October 2009
5. *Soviet Weekly* newspaper article, "Leonid Brezhnev and Fidel Castro congratulate Soviet-Cuban cosmonauts," issue 2 October 1980
6. Speech by President Fidel Castro at commemoration of Soviet-Cuban joint space flight at Havana City's Palace of Conventions, Cuba, 15 October 1980
7. Soyuz-38 experiments, taken from the site http://*www.prensa-latina.cu*

9

From the steppes of Mongolia

The son of livestock herder and cattle breeder Baldangiin Zhugderdemidiyn (1924–1998) and Chultemiin Ichinkhorol (1930–2002), Zhugderdemidiyn Gurragchaa was born in a traditional Mongolian portable dwelling known as a yurt on 5 December 1947 in Rashant, Gurvanbulag (Gurvan-Builak settlement) in the Bulgan province of Mongolia. He was the eldest son in a large family whose father never went to school, and whose mother had only four years of schooling. He saddled his first horse when he was still a child.

Zhugderdemidiyn finished his secondary education at a boarding school in the town of Bulgan and then attended the Ulaan Bataar Agriculture Institute in the Mongolian capital from 1966 to 1968. After eighteen months, however, he was drafted into the Mongolian People's Army and served as a wireless operator.

COSMONAUT CANDIDATES

In 1971, Gurragchaa traveled to the Soviet Union to study at a school for junior aviation specialists in an air base south of the town of Kant in the Kirgiz SSR, which is now in the Republic of Kyrgyzstan. He completed the course the following year and became qualified as a helicopter radio communications equipment mechanic. In July 1972 he entered the Zhukovskiy Air Force Engineering Academy, graduating in 1977 and returning to Mongolia. Also that year he married his girlfriend, Dashzevegeiin Batmunkh. They would later have two sons – Batbayar on 1 December 1977, and Odbayar on 7 July 1982 (the latter's name translates as "star holiday" and the birth came after Gurragchaa's space flight).

As an engineer back in Mongolia, and prior to his selection as a cosmonaut candidate, Gurragchaa was assigned to a separate squadron of the Mongolian People's Army as an engineer on aircraft equipment.

© Springer International Publishing Switzerland 2016
C. Burgess, B. Vis, *Interkosmos*, Springer Praxis Books, DOI 10.1007/978-3-319-24163-0_9

A young Zhugderdemidiyn Gurragchaa. (Photo: Author's collection)

Gurragchaa (second from left) in the Mongolian People's Army. (Photo: Author's collection)

Mongolia's cosmonauts: Maydarzhavyn Ganzorig (left) and Zhugderdemidiyn Gurragchaa. (Photos: Spacefacts.de)

The man who would become his back-up pilot, Maydarzhavyn Ganzorig, was born on 5 February 1949 as the son of a shepherd in the village of Ich-Chuzirt, located in the Cercerleg District of Mongolia's Ara-Changay Province. He took his early education in primary and secondary schools in Cercerleg, and later graduated as an automation engineer from the Order of Lenin Polytechnic Institute in Kiev, Ukraine, in February 1975, where his specialism was thermal energy systems. He then went to work at a power station in the Mongolian capital of Ulaan Bataar as a surveyor and engineer specializing in thermodynamics. In 1977 and 1978 he took on a position as scientist at the Mongolian Academy of Sciences Physical-Technical Institute, working on the automation of scientific instruments. The following year he joined the Mongolian Army Air Force as an engineer with the rank of captain. That same year he also became a member of the Mongolian People's Revolutionary Party.

TESTING TIMES

The selection process to find two suitable Interkosmos candidates from Mongolia began early in 1977, once the Soviets had approved the Special Committee's scheme for recruitment, its timing, and the medical and psychological requirements. In May, the flight records of all the country's military pilots were examined, and 30 were selected to undergo a full medical and psychological examination that month, although they were not told the purpose of these tests.

The testing apparently did not live up to the standards required by the heads of the Soviet Academy of Sciences. Three men had been selected, namely an engineer/pilot, Darzaagijn Syrenhorloo, and two pilots by the surnames of Saravsambuu and Erhambaar. These three were flown to Moscow for further examination. Shortly thereafter, Boris Petrov, Chairman of Interkosmos, arrived in Mongolia and advised the Special Committee that the position of cosmonaut was an extremely important one, and it was the feeling of his organization that the education of two of the candidates was not of the highest level. He requested that the search be extended to include civilian engineers and pilots. Of the group of three, only Syrenhorloo met the higher education requirements and was retained.

In October the recruitment process began once again, and this time it was not restricted to military personnel, allowing civilian engineers and even physicians to be considered. (The same situation would occur in the Romanian selection.) This time only 13 candidates were found suitable, and by the end of the year another three candidates had been selected to join Syrenhorloo in the final four:

- Darzaagijn Syrenhorloo
- Zhugderdemidiyn Gurragchaa
- Maydarzhavyn Gankhuyag
- Sanzaadambyn Sajncog.

Following a Mongolian Politburo directive issued on 12 January 1978, the four men were sent to Moscow for a final medical assessment at the Central Military Science Air Hospital (CVNIAG).

The four Mongolian candidates. At top: Maydarzhavyn Ganzorig and Zhugderdemidiyn Gurragchaa. Bottom row: Sanzaadambyn Sajncog and Darzaagijn Syrenhorloo. (Photo: Author's collection)

There was one problem of a rather unusual nature. Maidardzavyn Aleksandr Gankhuyag was told that he would have to change his surname to avoid offending or amusing Russian speakers. The surname Gankhuyag is a Mongolian word meaning 'armor', but the last two syllables resemble a Russian slang word for the male sexual organ. At the insistence of the Soviets he changed his surname to Ganzorig. (A similar issue occurred involving one of the Bulgarian candidates.)

The first candidate to be eliminated in the surveys was Sanzaadambyn Sajncog, when it was found that he had antigens present in his body that could have made him susceptible to inflammatory diseases. He was returned to Mongolia. Meanwhile, the other three finalists were approved. Since only two candidates could undergo the training program, another one had to be eliminated. The MPRP Politburo took the decision and in Directive No. 81 issued on 14 March 1978 it named the two finalists as Gurragchaa and Ganzorig and directed them to go to the Gagarin Cosmonaut Training Center near Moscow with their families.

Their cosmonaut training began in April 1978, first with tutoring in the Russian language and then moving on to techniques to be used in space travel. Later, they were given specific training in the systems and instrumentation in the Soyuz spacecraft.

As Gurragchaa later commented, "At the beginning. I was not sure that I would be able to cope with such sophisticated equipment, but Zvyozdnyy has excellent facilities for training and there were specialists who helped us a great deal in mastering space technology. When preparing for scientific experiments, we met scientists from various institutions and attended special lectures. We had to work rather hard at Zvyozdnyy."[1]

The four Soyuz-39 candidates stand in front of the Salyut training mock-up flanking the commander of the cosmonaut team Andriyan Nikolayev (center). From left: Vladimir Lyakhov, Maydarzhavyn Ganzorig, Vladimir Dzhanibekov and Zhugderdemidiyn Gurragchaa. (Photo: Author's collection)

Later the two Mongolians were assigned a Soviet commander. Gurragchaa was paired with Col. Vladimir Dzhanibekov, who had commanded the Soyuz-27 mission that docked with the Salyut-6 space station for a five-day visit in 1978. Ganzorig got Vladimir Lyakhov, who had commanded the Soyuz-32 mission in 1979, during which he and Valeriy Ryumin had docked with Salyut-6 and spent a record-breaking total of 175 days in space.

It was decided that Dzhanibekov and Gurragchaa would be the prime crew, with Lyakhov and Ganzorig acting as the back-up crew. This crewing was only openly confirmed two days prior to the launch.

Official portrait photo of the Soyuz-39 crew. (Photo: Spacefacts.de)

Gurragchaa and Dzhanibekov during water survival training. (Photo: Spacefacts.de)

FLIGHT INTO THE COSMOS

Prior to his son's flight, Gurragchaa's father Baldangiyn recalled for *Soviet Weekly* the early days of the boy's life. "When my son was small he rode in horse races. Of course I'm fond of him, and wanted him to become one of the best riders. Now he is putting his foot into the stirrup of a spaceship. I want him to honorably occupy his place among the worthy sons of our Earth."[2]

On the day of the scheduled launch, Soviet Academician Vladimir Kotelnikov, head of the Interkosmos Council under the USSR Academy of Sciences, told journalists that the majority of the experiments in the expedition's research program reflected the specific features of the Mongolian school of space research. In many cases the purpose was to study that country's natural resources, and this would involve laboratories acting simultaneously on three levels: aboard Salyut-6, on a special airplane, and on land. Cameras and spectrometers made in the USSR, the German Democratic Republic and Bulgaria would be employed to probe the areas of greater interest – Mongolia's ore-bearing lands, tectonic faults, high-altitude pastures, and promising oil and gas areas.

Professor B. Shirendyb, President of the Academy of Sciences of the Mongolian People's Republic, who was at the cosmodrome for the launch, said, "Our active participation in the Interkosmos program has given Mongolia the opportunity for the past fifteen years to develop the most modern fields of science – remote study of natural resources, space medicine and biology, and space communications. Those have had a direct impact on many areas of the economy, culture and health care."[3]

Dzhanibekov and Gurragchaa prepared for their mission in the Soyuz trainer. (Photo: Author's collection)

With the call sign of *Pamir* (a range of mountains), Soyuz-39 was launched from Pad 1, Site 5 at the Baykonur Cosmodrome on 22 March 1981; the pad from which Yuriy Gagarin had been launched almost 20 years earlier. The launch occurred just three weeks before the United States inaugurated a whole new generation of spacecraft with the maiden launch of space shuttle *Columbia*. The next day, Soyuz-39 rendezvoused with Salyut-6, orbiting 200 miles above the Earth. As the forward docking port was already occupied by the Soyuz T-4 spacecraft, they docked at the aft port thirty hours after lift-off. Soviet television beamed an image of Mongolian and Soviet officials leaping up, embracing and clapping at the moment the hard docking was confirmed. Television pictures then showed the new arrivals bobbing weightlessly into Salyut-6 to be warmly greeted by the residents Vladimir Kovalyonok and Viktor Savinykh. "Come on in, welcome," one of them urged, laughing and waving in a sequence watched by millions of Soviet citizens. Gurragchaa then released a bundle of red flags and they were seen floating in weightlessness.

After the ceremonies and being shown around the station by Kovalyonok and Savinykh, who had only been on board themselves for the previous ten days, Dzhanibekov mentioned that the station had undergone many changes since he had last been aboard in January 1978. But with some 30 experiments to carry out over the next eight days there was little time for acclimatizing or reminiscing.

WORKING ABOARD THE SPACE STATION

The main emphasis of the experiments planned by Mongolian scientists in cooperation with colleagues in the Soviet Union and other countries participating in the Interkosmos program, was laid on probing the Earth's natural resources and environment. The Erdem experiment, for instance, comprised three stages – photographing the territory of the Mongolian People's Republic from the Salyut station, photography and spectrometry from laboratory aircraft and investigations at ground level. The data the crew members accumulated would be useful to Mongolian scientists in studying the geological structure of mineralization zones, in mineral prospecting, mapping agricultural areas, forecasting crop yields, drawing up and finalizing soil charts, as well as in air and water pollution control.

The results of the Biosphere-Mon experiment would also prove to be of great benefit for the Mongolian economy. Its purpose was to study various objects on the Earth's surface by means of special instruments on Salyut-6. The crew carried out 14 observation assignments over the territory of Mongolia in the interests of geology, agriculture, glaciology, landscape science, and meteorology. Areas chosen for observation were:

- areas of seismic activity
- areas which might contain oil and gas reserves
- atmospheric processes
- environmental pollution
- underground fresh water reserves
- the Gobi desert and steppe areas.

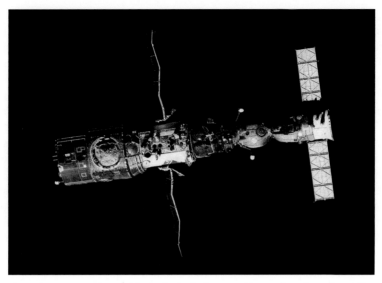

The Salyut-6 space station with a docked Soyuz spacecraft, taken from Soyuz-39. (Photo: Spacefacts.de)

The Atmosphere, Horizon, and Terminator experiments provided information about the transfer functions and other optical properties of the Earth's atmosphere. These were done using the Spectrum-15 instrument developed by Bulgarian specialists. This instrument was also used for the Illuminator experiment, which was a qualitative assessment of the changes in the spectral transparency of the Salyut station's viewports over its long orbital flight, and the Solongo experiment that aimed to obtain spectral data on various terrestrial formations.

Sometimes with the assistance of Kovalyonok and Savinykh, the international crew also conducted research into materials science. The Altay-1 experiment was to afford a deeper insight into the process of mutual diffusion in pure metals in microgravity, in this case lead and tin. The Altay-2 experiment grew crystals of vanadium pentoxide in the Splav furnace. These crystals are active semiconductors used not only in electronics, but also as efficient catalysts in obtaining many organic compounds.

Dzhanibekov and Gurragchaa carried out the Hologram experiment, which involved the development of new video recording and transmissions techniques which might one day be used to send holograms to and from Earth. The equipment for this had been developed by Soviet and Cuban scientists. As authors Brian Harvey and Olga Zakutnyaya note in *Russian Space Probes: Scientific Discoveries and Future Missions*, the purpose of the hologram test was "to transmit three-dimensional images of materials processing experiments, where there was a real value in getting a three-dimensional picture. Here, on 27th March 1981, the first ever holograms were transmitted from Salyut-6 to the ground using the KGA-1, a LG-78 helium-neon gas laser …. On that day, cosmonauts Vladimir Kovalyonok and Viktor Savinykh used the hologram in connection with the materials processing experiment, sending to the ground holograms of a sodium chloride crystal dissolving in water and then images of the window of the space station to assess impacts and microscopic imperfections."[4]

Using equipment developed by Soviet and Mongolian scientists, the cosmonauts carried out experimental research into primary cosmic radiation; the first experiment of this kind to be performed aboard the orbiting complex. The space particles were recorded by dielectric detectors mounted in the command module and airlock chamber. These recordings would enable specialists to determine the charges and energy levels of space atomic nuclei.

The crew also performed medical research into specific aspects of the human organism's adaptation to microgravity. In fact, this had involved collecting comparison data prior to the mission. The Biorhythm experiment, prepared by Mongolian doctors, showed the effects of the stress involved in this adaptation on the cosmonauts' biological cycles. This was of great practical importance because biological cycles have a direct bearing on one's work capacity. By knowing a cosmonaut's biological cycle their work capacity could be predicted and their work or experiments planned accordingly. Tables of the most important medical parameters were compiled on both men ahead of the mission, recording such data as temperature, blood pressure, and heart function. Readings were taken every two hours. An electrothermometer, alarm clock, and device for measuring blood pressure were used while aboard Salyut-6. The measurements showed that the indicators taken in space didn't differ appreciably from those obtained prior to the flight. Gurragchaa's temperature varied between 35°C and 36.2°C, his pulse between 62 and 68 beats per minute, and his blood pressure was in the range 117 to 124 millimeters maximal and 78 to 84 millimeters minimal.

There was also an experiment directed at the largely unknown scourge of space adaptation syndrome – the so-called 'space sickness' suffered by many space travelers from all nations. Because blood tends to flow to the head upon entering weightless conditions, crew members often suffered from headaches, light-headedness, and become vertiginous. This ailment was often exacerbated by sudden head movements. The experiment involved Dzhanibekov and Gurragchaa wearing a device which resembled a collar on their necks. This kept their heads from turning too fast. For the first three days on board Salyut-6 they would not remove the collar, and on the sixth and eighth days they would wear it for just thirty minutes.

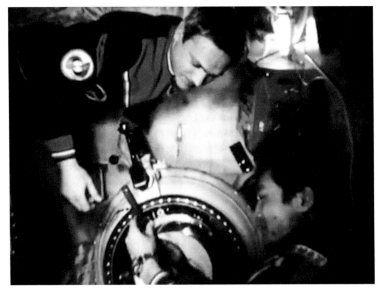

Dzhanibekov and Gurragchaa aboard Salyut-6. (Photo: Author's collection)

For the Blood Circulation-Sprint experiment, the cosmonauts were to observe the effects of blood redistribution on the state of the cardiovascular and respiratory systems during the period of "acute" adaptation to weightlessness, particularly during five-minute workouts on the station's Vel-3 bicycle training device (cyclergometer). The information thus obtained would add to the efficiency of monitoring a cosmonaut's state of health and help to develop better and more individualized use of onboard prophylactics.

Dzhanibekov and Gurragchaa continued the Soviet-Cuban crew's medical experiments. The Chatsargana experiment prepared by Mongolian scientists was of great value to space medicine. This investigated how adding sea buckthorn to food could alter certain aspects of metabolism in space flight. Sea buckthorn abounds in various biologically active substances and can therefore be used as a source of vitamin-rich foods. The findings of this experiment could also help to create alimentary preparations for increasing a person's adaptability to the highly demanding conditions of space flight.[5]

THE JOURNEY HOME

All of the experiments had been completed on 29 March, and after a period of relaxation the four cosmonauts began preparations to return the Soyuz-39 spacecraft to Earth, packing it to capacity with the results of the Soviet-Mongolian work and some of the experiments carried out by Kovalyonok and Savinykh.

At 12:57 p.m. Moscow time on 30 March, the heavily laden Soyuz-39 undocked from the Soyuz T-4/Salyut-6 complex and slowly withdrew. A soft landing at 2:41 p.m. brought to an end a successful mission for Dzhanibekov and Gurragchaa which had lasted 7 days, 20 hours, 42 minutes and 3 seconds.

There was a light fog and drizzle in the Kazakh steppes when the capsule touched down around 107 miles south of Dzhezkazgan, but the search-and-rescue helicopter group in the designated landing zone were able to quickly spot it. They caught sight of it as soon as the huge (10,000 square feet) orange parachute opened and kept it in view all the way down to the ground. The helicopters touched down at almost the same time.

On 30 March, Vladimir Dzhanibekov, who had already been awarded Hero of the Soviet Union after commanding Soyuz-27 in 1978, was awarded the Order of Lenin and a second Gold Star Medal. Zhugderdemidiyn Gurragchaa was awarded the title of Hero of the Soviet Union, the Order of Lenin, and the Gold Star Medal. The Presidium of the Great People's Khural of Mongolia conferred the title of Hero of the Mongolian People's Republic on both men, and also awarded each the Sukhe Bator Order and the Gold Star Medal of Mongolia. Upon their return to Earth later in the year, Kovalyonok and Savinykh were given the same awards for their cooperation in completing the Mongolian experiments on Salyut-6.

There have been persistent rumors that Gurragchaa became ill in space, possibly even incapacitated, but as with earlier Interkosmos flights very few photographs have ever been made public and very little information apart from that of propaganda value. According to authors Tim Furniss and David J. Shayler, "Gurragchaa may have been one of the few space travelers to have reacted violently to weightlessness. Only one photograph of him aboard Salyut-6 has ever been released."[6] However, in a video that shows Gurragchaa working on the station he appears to be in good health and spirits.

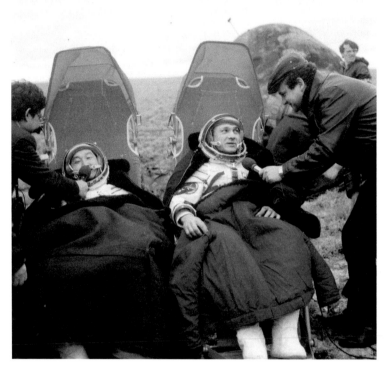

Gurragchaa and Dzhanibekov being interviewed after landing. (Photo: Author's collection)

LATER LIFE

Following his one and only space flight, Gurragchaa went on to become Chief of Staff of Air Defense for the Mongolian Armed Forces, and was promoted to the rank of major general. In 1988 he published the autobiographical book *To a Friendly World*. He then worked as chief of a scientific institute in Ulaan Bataar and later served as Mongolia's Defense Minister from 2000 to 2004.

Because he came from a nomadic family there was no real place of birth registered for the Mongolian space hero. This was solved by building a village on the place where he was born, which was named Kosmos.

As if his name wasn't enough of a tongue-twister for Westerners, in 2004 the 58-year-old Gurragchaa was obliged to select a new surname. This was because for more than 80 years everyone in Mongolia had been on a first-name basis. After seizing power in the early 1920s the Mongolian Communists had abolished surnames in their campaign to eliminate the clan system that was the hereditary aristocracy and the class structure. This meant that within a few decades, most Mongolians had forgotten their ancestral names, and used only a single given name. This system eventually became utterly confusing when 9,000 women ended up with the name *Altantsetseg*, meaning 'golden flower'.

By the mid-1990s, Mongolia had become a democracy again, and there were growing worries about the lack of surnames. One name might be sufficient when most people were nomadic herdsman in remote pastures, but now the country was urbanizing. The one-name system was so confusing that people were marrying without realizing they were related. A law passed in 1977 required everyone to have a surname and a system of citizenship cards was introduced. Slowly the country of 2.5 million people began to adopt surnames. In late 2004 the people of Mongolia were forced by the heavy-handed government to select a new surname, with anyone who failed to comply risking a financial penalty. When Gurragchaa was unable to discover his ancestral surname, he chose *Sansar*, the Mongolian word for the cosmos. His two children will use the same name.

A recent photograph of Gurragchaa. (Photo: video.xopom.com)

In March 2005, Gurragchaa was elected Deputy of the State Great Hural [Parliament] of Mongolia from the Mongolian People's Revolutionary Party, and he is also President of the Union of Mongolian Studies for Friendship with the CIS (Commonwealth of Independent States) Countries.

Ganzorig (left) and Gurragchaa in dress uniform. (Photo: Author's collection)

Gurragchaa and Ganzorig at the 30th anniversary awards. (Photo: Office of the President of Mongolia)

Following the Soyuz-39 mission, Capt. Maydarzhavyn Ganzorig worked in the Space Research Institute of the Soviet Academy of Sciences, where he defended a dissertation and was awarded a Candidate of Technical Sciences degree. In 1984 he decided upon a scientific career and was appointed head of an Image Processing Laboratory at the Institute of Physics and Technology at the Mongolian Academy of Sciences in Ulaan Bataar. Since 1991 he has been Director of Informatics at the Remote Sensing Center, also at the Mongolian Academy of Sciences. Married to a schoolteacher, with whom he has two sons, he is a member of the American Geographical Society and won its Miller Gold Medal for geodesy and cartography in 1997.

Zhugderdemidiyn Gurragchaa proudly displays "his" Soyuz-39 spacecraft in a purpose-built annex to his home. (Photos: Author's collection)

The Soyuz-39 descent module has had an interesting post-flight history. It was originally housed in a transportation museum in Ulaan Baatar during the 1980s, but with the fall of the Soviet regime in 1990 the museum was gutted and converted into a shopping mall.

The Mongolian Ministry of Defense was asked to remove the capsule. Quite remarkably this historic relic ended up securely housed in a specially constructed annex to the home of the Minister of Defense, who at the time just happened to be Zhugderdemidiyn Gurragchaa. In June 2014 the Space Center of Mongolia was opened in Ulaan Bataar and the Soyuz-39 capsule was finally placed back on public display along with an exhibition of equipment relating to the mission, including Gurragchaa's space suit.

REFERENCES

1. Rex Hall, David J. Shayler and Bert Vis, *Russia's Cosmonauts: Inside the Yuri Gagarin Training Center*, Praxis Publications, Chichester, U.K., 2005
2. *Soviet Weekly* newspaper article, "The Son of a Peasant Goes Up in Space," issue 2 April 1981
3. *Soviet Weekly* newspaper article, "Mongolia Joins the Space Club," issue 2 April 1981
4. Brian Harvey and Olga Zakutnyaya, *Russian Space Probes: Scientific Discoveries and Future Missions*, Praxis Publications, Chichester, U.K., 2001
5. G. Ryabov, article, "Soviet-Mongolian Space Flight Completed," *Soviet Weekly* newspaper, issue 9 April 1981
6. Tim Furniss and David J. Shayler, *Praxis Manned Space Log 1961–2006*, Praxis Publications, Chichester, U.K., 2007

10

Romania continues the program

Better known to his family and friends by his second name, Dorin, Dumitru-Dorin Prunariu is the son of an automobile engineer and a lady teacher. He was born on 27 September 1952 in the medieval city of Brasov, in Central Romania. During primary and secondary school he developed an early passion for flying machines, and was designing and building model airplanes along with other aviation enthusiasts from the Pioneer House in Brasov. During high school, together with a few colleagues, he built model rockets as a member of the 'Cosmos Messengers' Club, organized by a passionate teacher of his school. "Since I was a little boy I have had the desire to fly," he once stated. "I would shut my eyes and feel as if floating over the mountains and valleys, and discover new lands. The blue abyss and the infinite heights have always captivated me."[1] In a national contest, he was awarded a Bronze Medal for the design and construction of a rocket launching facility, and a Golden Medal for the design of a Classroom for Future Education. He then joined the 'Minitehnicus' National Club and received membership card no 103. Coincidentally, 11 years later he became the world's 103rd cosmonaut.

"I SAW ANOTHER FUTURE"

After graduating from a physics and mathematics high school in 1971, Prunariu decided to pursue his interest and to go into aircraft designing. He subsequently enrolled in the Faculty of Aerospace Engineering of the Bucharest Polytechnic Institute (which is now University 'Polytechnica' Bucharest), where he would meet his future wife, Crina, a faculty colleague.

Whilst an undergraduate he became a member of a flying club and took lessons as a sport pilot. Following his graduation in 1976 with the equivalent of a master's degree he worked as an engineer-in-training at the Brasov aircraft works in Ghimbav. In March 1977, he was sent to the Reserve Aviation Officers School in Bacau to perform his compulsory six-month military service. Upon completion, he received the rank of second lieutenant in the reserve.

While studying in Bacau, the commander of his unit told the students that Romania was about to participate in manned space flights and that candidates were being sought as future cosmonauts. Such a prospect seemed at that time remote and unreachable for Prunariu. He already had his career mapped out as an aviation engineer.

© Springer International Publishing Switzerland 2016
C. Burgess, B. Vis, *Interkosmos*, Springer Praxis Books, DOI 10.1007/978-3-319-24163-0_10

As Prunariu explained to the authors, "I only wanted to accomplish my military service, which was compulsory for all healthy young men in Romania in those times, and to return to the factory I was working with and to my family. For all higher education graduates, the compulsory military service lasted six months and it was performed in the reserve officers' school where we were supposed to learn everything about the aircrafts of our army and their operation. I was reluctant to the proposal of the commander to take part in the selection of candidates for the cosmonauts program because I thought my professional future was already defined. I knew little about Romania's involvement in space activities, even though Romania had been a part of the Interkosmos program since 1968, and we had already launched about seventeen different devices on board Russian rockets in the framework of the [Interkosmos] program. The commander therefore briefly explained to us about Romania's involvement in space activities and the latest level of cooperation within which participating countries were supposed to send their own candidate cosmonauts onto an orbital station where they would perform complex scientific experiments.

"We were pretty skeptical, but the commander said, 'Okay, [whoever] wants to take part in the selection will be taken to Bucharest by a military airplane.' I liked very much to fly, so I said, 'Oh, it's an interesting idea.' Secondly, he said, 'You will undergo a full medical check-up … for free, in the military hospital.' The military hospital was very well-known, a very professional hospital in that time, and we said, 'Why not? Maybe at this age, we have to think about our health.' I was only 25 years old. I was already married. We had one child, and my wife was in the last year of undergraduate study at the same faculty in Bucharest as I graduated from."

A portent of the future: a group of students (Prunariu at left) with a display of rockets. (Photo: Dumitru Prunariu)

Prunariu as a student pilot at the Brasov Aeroclub, 1975 (Photo: Dumitru Prunariu)

Prunariu found the offer of a flight aboard the military airplane to Bucharest and the free medical check-up very enticing, as well as the prospect of seeing his wife for several days, instead of staying in the military unit and marching for kilometers every day as part of his routine training program; he therefore raised his hand to volunteer.

INITIAL REJECTION

"It was somewhere in April 1977 when seventeen of us went to Bucharest. We flew on an Antonov-26, a military airplane … no seats inside, just benches for parachuting. We flew at 600 meters, it was noon, and some turbulence occurred. Some of our colleagues were very disturbed by the turbulence and threw up. When we arrived in Bucharest they said, 'No, it's not for us, this thing!'"[2]

Prunariu underwent the first medical check, with extremely tough parameters and norms imposed by the Soviets; higher even than those for pilots of supersonic airplanes. The entire group, Prunariu included, was rejected. The regular daily program and the good food in the military unit made the previously slim Prunariu put on some extra kilograms that his body had yet to process. This and a slight flu led to medical results that were unacceptable for a cosmonaut candidate. After a few days spent in Bucharest, the candidates returned to their military unit and soon they had almost forgotten about the tests. Four months later, five of them, including Prunariu, were sent again to Bucharest in order to repeat the tests, and this time he passed the examination.

The selection process of the Romanian cosmonaut candidates was accomplished by Air Force Command. It started in the spring of 1977 and included several levels. First, were the military engineers, then the civilian engineers who were doing compulsory military service (Prunariu was in this category), then the engineers working in various aviation factories. By September 1977, seven candidates were selected, among which only one was an active duty officer.

Two men were eliminated during the next training stage, and two gave up. Eventually, at the beginning of 1978, there were three of them left: Capt.-Eng. Mitica Dediu (whose name was later changed from Mitica into Dumitru), and the two civilian engineers Dumitru-Dorin Prunariu and Cristian Guran. Prunariu and Guran were transferred from their factories to the military unit in Bacau to begin the preliminary training program organized by the Romanian Military Aviation. Dediu was already an active duty officer with that military unit.

By the completion of the selection of the two candidates who were supposed to go to Star City near Moscow, the three candidates had been trained initially in Bacau, then in Poiana Brasov where the Military Aviation had some facilities, and then in the Military Academy in Bucharest. In January 1978 they were sent for a final medical examination in Moscow. The training included a course in the Russian language, but the proficiency they achieved hardly allowed them to express themselves in elementary Russian and to understand the basic ideas of what was spoken.

Following the tests in Moscow, Guran was eliminated due to problems with his vestibular system. Only Dediu and Prunariu remained to accomplish the entire candidate cosmonaut training program. However, the commander of the Romanian Military Aviation considered that Guran was worth training and asked him to remain a back-up active duty officer with the Military Aviation. It was further proposed that Prunariu and Guran become aviation active duty officers. Prunariu had never considered a career in the military. He was married and he already had two children at the end of December 1977, but the offer was almost too attractive to decline. On 13 March 1978 Prunariu and Guran were both promoted to the rank of senior lieutenant, while Dediu was promoted to the rank of major "for extraordinary performance of duties."[3]

On 20 March 1978 the two Romanian candidates set off for the Yuriy Gagarin Cosmonaut Training Center near Moscow to complete their training and increase their proficiency in the Russian language.

The selection of the Romanian candidates, and particularly Prunariu's selection, offers something of a surprise when it is understood that most of the other Soviet bloc prime and back-up international cosmonauts were military pilots, whilst Prunariu, for instance, was initially an aerospace civilian engineer. Mongolia also sent officer engineers who were graduates of the USSR higher education. As Prunariu explained, "Russia's only proposal was that countries should send people who had studied in Russia … because of the language proficiency requirement. But Romania … you know, was politically a little bit different than the other socialist countries. Starting with 1964, we stopped sending people to any military institutions in Russia. So, at that time, we didn't have anyone speaking Russian and having graduated from Russian military or civilian institutions. So we had to learn Russian. The second issue was that Romania imposed on the candidates the requirement to be engineers … to have a technical higher education degree, to be able to work with the advanced level of space technology."

Cristian Guran, having the background of an electronics engineer, became an expert in air navigation and held several positions in the Romanian Military Aviation. Towards the end of his military career, he held the rank of air fleet general and worked for a time as Romania's representative with EUROCONTROL, which is the European Organization for the Safety of Air Navigation. Then after becoming a reservist, he returned to that organization as a private expert.[4]

Romanian cosmonaut candidates Dumitru Dediu (left) and Dumitru Prunariu. (Photos: Spacefacts.de)

STAR CITY

Having arrived in Star City to begin their training in March 1978, Dediu and Prunariu found that the first few weeks were devoted to lessons on space technologies, spaceship dynamics, and rules about space travel, with lessons taking up to eight hours per day. It soon became apparent that their low proficiency in Russian would seriously hinder this training, so it was decided to bring in professional teachers. Aleksey Leonov was the commander of the Soviet cosmonaut team at the time and he made all the arrangements with the 'Patrice Lumumba' University, which was an institution dedicated to teaching Russian to foreign students.

"We were fortunate also to have a very good lady teacher," Prunariu recalled. "She was around her fifties at that time, with a high level of general knowledge. She announced from the very beginning that we would not be using dictionaries and we [would] have to learn to think in Russian … because if we used a dictionary, we instantly made a connection between our language and Russian. If we didn't use it, and hence had in mind only the explanation in Russian of that word, we'd start to think in Russian. And she tried in a very efficient way to explain to us all new words, expressions, everything … And step by step, we developed our skills. But we had to learn a lot at home. In half a year, I could speak pretty good Russian and understand, let us say, maybe about 90 percent of conversations. Altogether, we learned Russian for a year and a half. During the language learning process, we worked on Russian literature, we were analyzing very complex texts, and we were studying a lot of grammar. I can say that after one and a half years of studying Russian dutifully and living in a Russian environment, one could hardly tell the difference between me and native Russians."[5]

Dumitru with his family in Star City. (Photo: Author's collection)

Prunariu undertakes weightless training in a specially modified aircraft. (Photo: Author's collection)

It was not long before the two candidates received Soviet mission commanders for further training purposes. Prunariu was crewed with the veteran cosmonaut Yevgeniy Khrunov, who had flown on the Soyuz-5 mission in 1969, while Dediu was to train with Yuriy Romanenko, who had recently completed an Interkosmos mission along with the

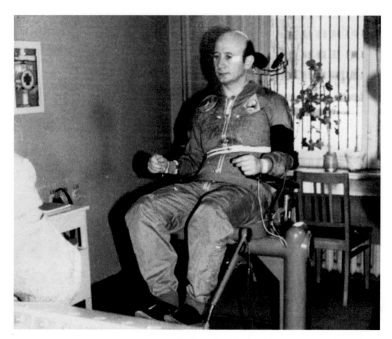

Dediu undergoing medical tests in Star City. (Photo: Author's collection)

Cuban Arnaldo Tamayo Méndez. However, Khrunov unexpectedly was forced to resign from the cosmonaut team on 25 December 1980. It was heavily rumored that he had been asked to resign for a mysterious "violation of regulations".[6] While not going into details, Prunariu hinted that this was due to marital issues ("Some trouble with his wife," he stated), which the Soviet space hierarchical authority structure evidently found unacceptable. This news came as a surprise to Prunariu, and it was several weeks before he was assigned a new commander.

"For about two months I had to train by myself," Prunariu said, "accomplishing, in the simulator, the job of the researcher, or space engineer, and of the commander as well. It helped me to better understand the role and the responsibilities of the commander, and the buttons that the commander must press aboard the spaceship at certain times. I had a young and ambitious trainer, Nikolay Chukhlov, who became my friend. Together, we were trying to define creative, innovative solutions in my training, which was increasingly appreciated. Nikolay had graduated from an aviation institute with the best results, and he had become a trainer of cosmonauts. He was a bit younger than me and I was his first student. We spoke the same language of engineering; I was acquiring very fast the knowledge he was sharing, and I was even able to find optimized solutions to implement activities in the simulator. So we were a very good team, teacher and student. The fact that we became friends helped the process. In the summer of 1980, Leonid Popov came back from a long flight on Salyut-6 … He spent 186 days aboard the station. After several months, I think the end of November, I cannot recall exactly the date, he was proposed to replace Khrunov and to start the training with me.

Following Khrunov's replacement in Soyuz-40 training, the two crews became (at top): Yuriy Romanenko and Dumitru Dediu and (bottom) Leonid Popov and Dumitru Prunariu. (Photo: Spacefacts.de)

"While I trained with Khrunov, we had a lot of problems; not of language, but of mutual feelings towards each other. He was much older, he hadn't trained for a very long time, and he was always stressed and nervous during the sessions. When Leonid Popov came along, everything became very easy. He was very experienced, he had flown half a year, he knew perfectly how to react and what to do under various circumstances in the Soyuz spacecraft and in the Salyut laboratory, but he didn't know very well the theory. It is what I was very good at, therefore we mutually decided how to share responsibilities during the training and examinations to have the maximum score. He proposed to me, 'If we receive questions from the trainers [who were outside of the simulator], you answer the theoretical questions, and I do the practical work.' We made a deal, and it worked perfectly well. We got along well with each other, we had a good compatibility.

"Bottom line, during the final examinations before the space flight, we had the maximum score ever obtained in Star City by any crew."[7]

THE FINAL CHOICE

Throughout their training, assessments were carried out on the two crews and it soon became obvious that the Popov-Prunariu team provided a stronger combination and they were given the nod as the prime crew, although this was not made official until shortly before the launch date. Despite this, Prunariu knew that his qualifications and examination marks were rated high, and rumors from various sources assured him that he would fly the mission.

Training with their respective mission commanders in the Soyuz simulator. At top, Dediu and Romanenko; at bottom Prunariu and Popov. (Photos: Spacefacts.de)

"After the final exams, I had only fives, even more than any of the Russians. Being an aerospace engineer, the theory, the construction, and the aerodynamics of the spaceship was something very simple to understand for me … and also having some flight experience as a sport pilot during my [earlier] studies. I managed to drive the spaceship in a slightly more skilled way. After I started training, I became a fan of cosmonautics. I was curious about everything connected with it; I read a lot; I learned a lot."[8]

The four Soyuz-40 crewmembers pay tribute to Yuriy Gagarin. (Photo: Spacefacts.de)

Asked about the choice between himself and Dediu, Prunariu said that the final decision belonged to the Romanian authorities, and it was based on their training outcomes. He said that he trained at the same time as the Cuban candidate Arnaldo Tamayo Méndez, where the choice was left up to the Cuban authorities, and he felt there were far more political issues associated with the selection of Méndez.

"Both candidates were very well trained, very well prepared, but Tamayo was the first black person to be sent into space. That was a political challenge for Russia … for the Soviet Union in that time, to send him. He was well prepared, no doubt, but his Russian was pretty poor, compared to the other one [José López Falcón]. Also a political choice was done with the Vietnamese team, because [Pham Tuan] who was sent into space was already a Hero of Vietnam, [having shot] down a B-52. But the second one [Bui Thanh Liem] was also very well trained and he spoke very good Russian. Unfortunately, he died in a military airplane shortly after returning home. For Romania, and especially taking into account the sensitive relations between Romania and Russia, the Russians left it to the Romanian side to decide. And the Romanian side took the decision."[9]

Although the names of the two candidates and their respective flight commanders had been revealed some time earlier, there was an embargo on releasing the names of the prime crew, presumably in case there was a last-minute problem with any of the four men. This embargo was lifted two days before the flight was due to launch, and the names of Leonid Popov and Dumitru Prunariu were formally announced as the prime crew. Their back-ups were Yuriy Romanenko and Dumitru Dediu. It was a bittersweet day for Dediu, since the crewing announcement was released on his birthday.

Prunariu undertaking egress and water survival training accompanied by Hungarian candidate Bertalan Farkas. (Photo: Author's collection)

Popov and Prunariu answer questions prior to their Soyuz-40 mission. (Photo: Spacefacts.de)

A ROMANIAN IN SPACE

On 14 May 1981, the Soyuz-40 spacecraft was launched into darkness at 9:17 p.m. Moscow time from the Baykonur Cosmodrome in Central Asia, carrying Leonid Ivanovich Popov and Dumitru Prunariu. With this flight, Romania became the ninth country of the Soviet bloc to send a man into space aboard a Soviet spacecraft.

According to the Interkosmos flight planning, their mission would last for about 8 days, until 22 May 1981. After docking with the Salyut-6 space station on the second day, they joined Soyuz T-4 cosmonauts Vladimir Kovalyonok and Viktor Savinykh, who had manned the station since 12 March. After resting, Popov and Prunariu began their research program, which included medical and biological studies and experiments of specific scientific interest to Romania. The topics included astrophysical experiments, study of the radiation environment, biomedical experiments, psychological tests, and technological experiments. In fact, all four crewmembers would assist in completing a total of 22 scientific experiments for this phase of the Interkosmos program.

It is known that cosmonauts smuggled all sorts of things on board. Cameras, candy, even alcohol have covertly made their way into orbit. Despite the fact that they signed a document stating they wouldn't fly alcohol or a camera, Prunariu and Popov carried aboard Soyuz-40 a small, automatic Canon camera with a 38-mm lens that French candidate Patrick Baudry had purchased for Popov for around $100. Prunariu said that the quality of the pictures this took was better than those taken with the 24-mm and 50-mm Praktica and the 80-mm Hasselblad cameras that were aboard as the official equipment. The advantage of the Canon was that it didn't need a lot of time to set up. He added that getting the camera on board the Soyuz had been easier than he had anticipated. In fact, he said, he and Popov had had some inadvertent help from rocket engineer and designer Valentin Glushko.

Three days before the flight the crew had received a package which was wrapped in paper, tied with a green ribbon and marked "Cleared for Flight". The parcel was signed by Glushko, and held A4-size diplomas which the four men were to stamp and sign while in orbit aboard the Salyut space station. These were later to be given to some important leaders as proof that the Soyuz-40 crew had been on board the station and had accomplished space experiments together. However it provided them with an excellent chance to secretly add some personal items. Popov, helped by the cooks, made a cardboard box the size of the diplomas that was sufficiently thick to accommodate the camera, several rolls of Kodak film, and a 200 ml flat bottle of Romanian brandy. Popov and Prunariu carefully unwrapped the package supplied by Glushko, added their box, and repacked the parcel using the same wrapping and ribbon. Now the package was thicker than before, but since it was marked "Cleared for Flight" and signed by Glushko, nobody had the nerve to question its contents.

The Canon camera was used extensively to shoot souvenir pictures in orbit. Upon their return to Earth, however, Popov and Prunariu had no option than to hand the flight films to the Zvyozdnyy photo lab for development. According to Prunariu, it instantly became clear to the lab technicians that this was not material shot with the official equipment. He said so many photos had been made during the earlier missions that they knew exactly what would be in a frame taken with cameras using the 'official' 24-mm and 50-mm lenses.

The journey begins for the crew of Soyuz-40. (Photo: Spacefacts.de)

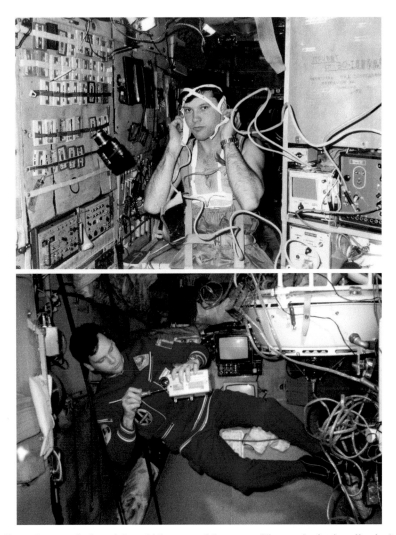

Prunariu at work aboard the orbiting space laboratory. (Photos: Author's collection)

As the crew commander Popov was asked how this could be, being urged to tell, with the promise that his explanation wouldn't be repeated to anyone else, as these photographs were by far the best of the Interkosmos program and they wanted to use them. Finally, their secret was revealed.

As Prunariu revealed, "Popov told me that, 'You know, the specialists from Star City have seen that the focal length of that lens was not that of Praktica.' And they asked, 'If you don't say what you carried up there and what you did, we may punish you, and so on.

Prunariu and Salyut-6 crewmember Viktor Savinykh with newspapers the Interkosmos crew had delivered to the space station. (Photo: Author's collection)

But if you tell us, we won't say anything and we will publish the pictures, because they are some of the best ones taken on board with international crews.' Popov admitted, 'Okay, we had a small camera, we took these pictures.' They were published as official pictures, taken with official cameras on board. Nobody knew that we had used another camera. But always a specialist, having many images from the space station, can determine exactly the focal length of a lens used to take a picture. So, they appreciated the Canon camera was not on board … officially. So, this was the story." Despite their cooperation, Prunariu never saw the negatives again, although he does have prints taken from a number of them.[10]

In an interview for the *Soviet Weekly* newspaper, he was asked what the Earth looked like from space for him.

"In orbit, one gets a very clear view," he replied. "For example, at one glance, you can see a huge area [extending] from the Nile to the Caspian. And you see everything exceptionally well, even better than on a map – mountains, faults and ring structures. It is easy to get one's bearings and identify what territory you are over at that moment. Unfortunately, Europe was heavily overcast. And we passed over Romania in the evening – only on the final days of the flight did it come out of the shadow, enabling me to see the Danube and the Carpathians."[11]

Prunariu said that he was surprised – even shocked – to look down on a sea of lights while flying over the United States at night, and many lights in western European cities, but few in eastern Europe's largest cities, and almost complete darkness elsewhere in those countries.

Prunariu at a porthole window on Salyut-6. (Photo: Author's collection)

FALL FROM GRACE

Popov and Prunariu undocked from the Salyut-6/Soyuz T-4 complex at 2:37 p.m. Moscow time on 22 May and, 3 hours and 21 minutes later, landed safely about 140 miles south-east of Dzhezkazgan.

Popov and Prunariu perform the tradition of chalking their signatures on the Soyuz-40 descent module. (Photo: Author's collection)

On 2 June 1981 the General Secretary of the Central Committee of the Communist Party of the Soviet Union, Leonid Brezhnev, presented decorations to the Soyuz-40 crew. Leonid Popov received the Order of Lenin and a second Gold Star Medal. Dumitru Prunariu gained the title of Hero of the Soviet Union, the Order of Lenin, and the Gold Star Medal.

With great candor, Prunariu confirmed that following his return to Romania he had more or less fallen out of grace with the Ceausescu government, principally due to his popularity. "When I went to Star City in 1978, the situation wasn't as bad as when I came back in '81. I was not at all involved in politics [and] I didn't know exactly what happened from a political point of view. I had only been focused on my professional work at my level, and that was all. But when I returned, Ceausescu's wife [Elena] was much more involved in politics, and she was worse for the political situation in the country than Ceausescu. She influenced him in a lot of things … decisions connected with many activities in the country. And, of course, the leaders below them were obliged to do what she told them to.

"During our training in Star City in '79, the Russians invaded Afghanistan, and Romania was officially against this. In '68, when the Warsaw Treaty countries (with the exception of Romania) invaded Prague and the Czech Republic, Romania also officially declared against this practice. So the Russians looked at us with some sensitivity. Furthermore, Ceausescu looked at the Russians with some distance and sensitivity. But they, in a way, forgot that I was a Romanian military person, in Russia to accomplish a mission on behalf of Romania. That was my official status. When I came back I already had the title of Hero of the Soviet Union, given by Brezhnev."

In fact, when Prunariu arrived at Otopeni Airport following his space flight, neither the head of state nor the Prime Minster was present to greet him and his Russian colleagues, in contrast to what had earlier occurred in other countries of the Interkosmos program. When Aleksey Leonov saw this, he said to Prunariu, "Dumitru, your future looks bleak." Prunariu did not believe him at the time, but later began to realize that the Ceausescus would tolerate no other heroes.

"In that time, we had a Minister of National Defense who was very afraid of Ceausescu and couldn't say anything in front of him. He was always just behind, and never saying or proposing anything. Before meeting Ceausescu, I asked Olteanu if I should wear the Soviet order received from Brezhnev during the meeting, or if I should put it in my pocket and then wear it again after the meeting. I didn't say it, but it crossed my mind to give Ceausescu the impression that he gave me the only order, and at the same time not show him an award that he himself hadn't received. Ceausescu was very sensitive to decorations, and he very much liked to be the first to receive awards without doing anything. Olteanu said, 'You must wear it because the Russian side could be offended if you don't.' So I wore it. Ceausescu looked at me, looked at that award, and gave me the Romanian award.

"By this time [having spent so much time in training] I had a slight Russian accent. I had to read a speech they gave me in front of Ceausescu. It was written for me; I couldn't make any modifications. So I read it with my slight Russian accent. Ceausescu just stayed back a bit and didn't look like he was paying too much attention to me and to the crew.

"Just before the ceremony I heard Ceausescu's wife saying, 'Ah, you know … with this modern technology, it's so easy to fly into space.' I became angry. I couldn't say anything in front of her at the time, but in my mind I knew that she was not informed by her advisors

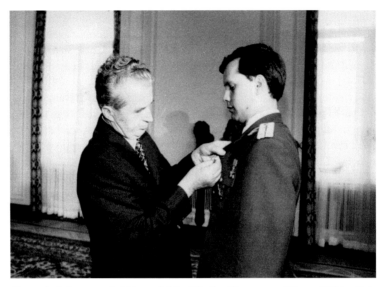

Prunariu being awarded his medal by Nicolae Ceausescu. (Photo: Wikipedia)

of what a space flight means; what sacrifices the cosmonauts make; how many have died in making a space flight … She didn't know anything about it. Even though she held a PhD degree in chemistry, she was not educated. Only after 1990 I found out that she graduated only four classes and got diplomas as a member of the Communist Party. Besides that there was another really political aspect [as] the people compared me with Nadia Comaneci, the well-known gymnast of Romania who in 1976 scored the maximum points in [the Olympic Games hosted by] Montreal [in Canada] and was famous all over the world. She received a Romanian order that was lower than mine. I was the Hero of Romania, she was the Hero of the Socialist Labor Order – also a very high one, but not as high as mine. But for Ceausescu, Nadia Comaneci was the image of the socialists, winning in front of a capitalist country. She was prepared in Romania, trained in Romania, and achieved success, the fame, in a capitalist country. Me, I was a Romanian officer, trained abroad using foreign technology, and flying into space on behalf of Romania with a country with which Romania had not very friendly relations.

"So, from a political point of view, Ceausescu didn't want to have a new hero in Romania. He was supposed to be the only Hero of Romania in that time, even if the population was fed up with him, was too much disturbed by his appearances on the television all day, and so on. The population needed a fresh, new face – innocent, let's say – to represent them somewhere else other than in politics."

Prunariu's space flight could have given him popularity well beyond that of Ceausescu's, but the leader and his minions deemed this unacceptable. "I could rise too high in their eyes, and they cut my appearances on television, completely. They never published stories about me on the first pages of the newspapers, just restricted me to a corner on the last page of the newspaper. I was never actually told about this; I just noticed that they never asked me for interviews. I was a military man with a good salary, having a good life, but being involved only partially in such things. And that was the situation."

Public appearances were few and Romanian television was forbidden to show Prunariu. The only film made by Romanian television about him, *Un Roman Zboara Spre Stele* (A Romanian flies to the stars), went into the archives without ever being shown.

Prunariu was asked if he was allowed to travel abroad following his space flight. "At the very beginning it was not so easy. For about one year I was kept at home. According to the military regulations in that time, if you spent more than three months abroad, you needed to remain at home for two years inside the country. After one year they made some exceptions because I was asked, as a public person, to take part in a lot of space activities abroad. I was invited to visit different countries, especially to Russia because the Academy of Sciences of Russia spent a lot of money organizing activities with foreign cosmonauts at that time. So I was allowed to go. I didn't disturb anybody, I never acted against my country, I never said anything; and I was allowed to take part in a lot of other activities elsewhere in the world."

Asked how the *Securitate* authorities reacted to the fan mail that Prunariu received from abroad asking for photographs and autographs, he said, "All these letters passed through the command of the air force. None came to me at home, because I was not allowed to give my home address at that time. Everything came through them, and the secretary just collected them and gave them to me to sign and they then sent them to the people who had asked for photographs and autographs. I got a lot of letters. No one was restricted, and the military command understood the situation and my position abroad."

BACK IN THE AIR

Until Ceausescu was deposed in 1989, Prunariu found himself in something of a limbo. No one, it seemed, knew how to utilize his experience as a specialist in space activities. Instead, they created for him a position within the Romanian Air Force, making him Main Inspector for Aerospace Activities.

Dumitru Prunariu with the Soyuz-40 descent module in 2011. (Photo: Nini Vasilesku)

"It was the highest position for an officer in that time, with a good salary," he recalled. "I was a captain and belonged to the flying staff. I already flew in Russia; I flew the L-39 [Aero Vodochony Albatross, a high-performance jet trainer developed in Czechoslovakia]. To keep the status of a flying staff, I was obliged to fly also in Romania. According to the regulations of the Air Force Command, if you don't fly for two years, you lose the title of flying staff, so the commander just let me fly together with an instructor [on the] L-39. I went to the military unit which had these airplanes, and time after time we gained some flying hours. After about one and a half years, the Minister of Defense found out that I was flying jets, and he insisted, 'No! You know what happened to Gagarin! We have to keep him. He's our hero.' And so he didn't permit me to fly jets any more. But the commander of the air forces said, 'Okay, I am not allowed to let you fly jets, but it is in my power to allow you to fly as a navigator on military transport airplanes.' So I took some lessons as a navigator and I continued to fly on the Antonov-26 and -24 as a navigator, just to keep my status on a flying staff, because this gave me the advantage of a higher salary, of some incentives connected with food, military clothes, and so on. Why not? The period was not so easy for Romanians, especially in that time before the '90s, when Ceausescu tried to pay back all the external debts of Romania – didn't import any goods for the population, just exported everything – while the population became poorer and poorer because of these efforts. Any advantage offered by officials was welcome and I accepted. I continued to fly, and kept my status in the air force until 1998.

"In 1998 I was a colonel, and I was asked by the Minister of Education and Research to move from the Ministry of Defense to his ministry as the President of the Romanian Space Agency. I always had been involved with the space agency from the very beginning, from the '90s, but from the outside. The air forces let me spend part of the working time within the Space Agency, let me be involved in different scientific trips abroad, and they included all these activities in my responsibilities as the Main Inspector of Aerospace Activities.

"Going back a bit, right after Ceausescu was deposed, the officials started to pay much more attention to me as a person who could represent something outstanding for Romania, inside and outside. When Ceausescu was deposed, he and his wife were tried by an ad hoc court and shot by a firing squad on the 25th of December 1989. After that I wrote together with my Romanian space friends a memorandum to the new government, headed by Petre Roman. We wanted to reorganize the space activities and space institutions in Romania in the framework of a new Space Agency. We suggested this could be a budgetary institution under the auspices of the Ministry of Education and Research. After two days, Petre Roman called me, 'Okay, I need you; come to me. I have news for you.' At that time I was a major in the Air Force Command, and I was very happy to think that he had read the memorandum and would maybe approve the organization of the new Space Agency."

It transpired that Roman had not read Prunariu's memorandum about creating the Space Agency; he actually wanted to make him chief of the Civil Aviation Department. Prunariu mentioned that he was a military officer, but Roman dismissed this by saying he could be moved from the Ministry of Defense to the Ministry of Transportation. Not only that, he would be offered the rank of general. Prunariu felt that was too high a jump in rank and not altogether fair on his colleagues in the Air Force Command. Eventually they settled on the rank of colonel, and the former cosmonaut-researcher found himself working over the next eighteen months as both Deputy Minister of Transportation and chief of the Civil Aviation Department.

"There were four successive Ministers of Transportation during that time. There were big social changes … trade union strikes … I had to face a lot of difficult situations and I learned a lot during that time. I even accomplished a fellowship in Montreal, Canada, at IAMTI, the International Aviation Management Training Institute, which is for civil aviation. My eyes were reopened as to what happened in the world and what we had to do to raise the level of international involvement of Romanian Civil Aviation. But the fourth minister, Mr. Traian Basescu, who later became president, receiving my proposals for improving the activity, just told me, 'I don't care about your aviation. I don't want to make any change in this. And if you insist, I will change you.' I was angry and I resubmitted the proposal document through the registration desk of the ministry. He was not used to being defied by anyone; he became angry as well, and the next day he changed me; I went back to Military Aviation. And so in 1998 I was appointed president of the Romanian Space Agency."[12]

The Soyuz-40 mission was the final space flight in the original Interkosmos series. To celebrate, several crewmembers gathered in front of the Salyut training simulator along with a number of veteran Soviet cosmonauts. (Photo: Author's collection)

POSITIONS OF EMINENCE

From that time on, Prunariu worked within the Space Agency. In 2004 he agreed to being appointed as the Ambassador of Romania in the Russian Federation. In the fall of 2004 the presidential elections in Romania were won by the former minister of transportation, Mr. Basescu, heading the opposition. In 2005 he paid two official visits to Russia, and he met Prunariu as the Ambassador. Although not having any reason to disapprove of Prunariu in professional terms, the simple fact that he did not belong to the new clique was enough for Basescu to replace him in the summer of 2005.

In 2004, when Prunariu left for Russia, he had resigned from the position of President of the Romanian Space Agency, being succeeded by the executive officer, who was his friend Marius Piso. From 2005–2006 Prunariu served as President of the Board of the

Romanian Space Agency. By mutual agreement between the Space Agency, the Ministry of Research and the Romanian government, in 2006 Prunariu was sent to the European Commission in Brussels, Belgium, to represent the minister as director of the Romanian Office for Science and Technology (ROST).

In February 2007 Prunariu finally retired from the Ministry of Defense with the rank of major general, albeit continuing to work as a civil servant. He continued in Brussels until September 2008. Then from 2008 to 2013, he was President and Member of the Board for the Scientific Council of the Romanian Space Agency.

In 2010 Prunariu was elected president of the European Chapter of the Association of Space Explorers (ASE), and the following year took over full presidency of the ASE for a period of three years; in 2014 he was re-elected for another term. Meanwhile, he had been elected Chairman of the United Nations Committee on the Peaceful Uses of Outer Space (COPUOS) for the period June 2010 to June 2012.

In November 2011 Prunariu became an Honorary Member of the Romanian Academy and in September 2012 he received the 'Social Sciences Award' of the International Academy of Astronautics for his significant and lasting contributions to the advancement of astronautical sciences. In October 2012, as a result of his professionalism, experience, and correctness, he was appointed to chair the Nominations Committee for electing officers to the International Astronautical Federation. Also, since in 2013 he has served as the President of the Romanian Association for Space Technology and Industry (ROMSPACE).

Prunariu is keen digital photographer, his PhD thesis dealt with new developments in the dynamics of space flight, he has presented and published numerous scientific papers and co-authored a number of books on the topic of space technology and space flight:

- *La cinci minute după cosmos* (Five minutes after a space flight), 1981, Military Publishing House, Romania
- *Cosmosul-Laborator şi uzină pentru viitorul omenirii* (Outer space – laboratory and factory for the future of humankind), 1984, Technical Publishing House, Romania
- *Istoria aviaţiei române* (History of the Romanian Aviation), 1984, Scientific and Encyclopedic Publishing House, Romania
- *Dimensiuni psihice ale zborului aerospaţial* (Psychological dimension of the airspace flight), 1985, Military Publishing House, Romania.

Over the years, he has accumulated a number of significant awards:

- 1981, Hero of Romania and Hero of the Soviet Union. Awarded for the successful accomplishment of the first Romanian-Russian space flight
- 1982, 'Yuriy Gagarin' medal of the International Astronautical Federation
- 1984, 'Hermann Oberth Gold Medal' awarded by the German Rocket Society 'Hermann Oberth – Wernher von Braun'
- 2000, Order 'Star of Romania' with High Officer Degree awarded by the President of Romania for his professional activity
- 2010, 'Aeronautical Virtue' Order awarded by the President of Romania
- 2011, 'Medal for Merits in Space Exploration' awarded by the President of the Russian Federation
- 2011, 'Academic Merit' Diploma of the Romanian Academy
- 2011, 'Emblem of Honor of the Romanian Army' awarded by the Ministry of Defense of Romania.

INTERKOSMOS
PROGRAM PATCHES

SOYUZ-28
(CZECHOSLAVAKIA)

SOYUZ-30
(POLAND)

SOYUZ-31
(GDR)

SOYUZ-33
(BULGARIA)

SOYUZ-36
(HUNGARY)

SOYUZ-37
(VIETNAM)

SOYUZ-38
(CUBA)

SOYUZ-39
(MONGOLIA)

SOYUZ-40
(ROMANIA)

Annie Muscă's updated biography *Dumitru-Dorin Prunariu: Biografia Unui Cosmonaut* was published in Bucharest by the Adevărul publishing house in May 2012. The following year Prunariu was saddened to read of the death of his good friend and former Interkosmos partner Dumitru Dediu, who retired from the Air Force in 1997 and died in Bucharest after a lengthy illness; he was laid to rest on 11 July 2013.

Today, Dumitru Prunariu and his wife Crina Rodica (born 6 January 1953) have two sons and three grandchildren. After Crina graduated from the Aerospace Engineering Faculty of the Polytechnic University of Bucharest, she worked for a short time in an aviation factory. Back from Russia after her husband's space flight she graduated from a master program in foreign trade and worked within a company trading aviation parts, then later as an expert in the Ministry of Foreign Trade. In 1990 she was moved to the Ministry of Foreign Affairs as a specialist in foreign trade. She later graduated from the Diplomatic Academy and became a career diplomat. Her last position was of the Ambassador Extraordinary and Plenipotentiary of Romania to Armenia from 2007 to 2013.

Their elder son Radu-Catalin was born on 28 December 1975 in Tirgu Mures of Mures County. He graduated the Dimitrie Cantemir National College in 1994 and was awarded a bachelor of arts degree in financial management in 2000.

In 1996 the Romanian postal services issued a stamp bearing Prunariu's portrait. It was designed by his son with the aid of a computer to commemorate the 15th anniversary of his space flight. Prunariu said at the time that he was the only living Romanian whose portrait could be found on a stamp. For a while, Radu-Catalin ran a web and software development company, and also provided IT consultancy services. In 1997 he earned his private pilot's license, and in 2003 his commercial pilot's license. He commanded the Aurel Vlaicu Air Club in Bucharest, graduated from the Civil Aviation Academy, became a flight inspector within the Civil Aviation Authority of Romania, and is now a commercial pilot within the national Romanian carrier TAROM, and a member of the Romanian aerobatic team.

Younger son Ovidiu-Daniel was born on 15 December 1977. He took university courses (British studies) at Budapest ELTE University from 1997–2001. Since 2001 he has been a student at Spiru Haret University in Bucharest, in the faculty of International Relations and European Studies. After working within private institutions he passed an exam to become a PR and communication adviser to the director of the Civil Aviation Authority of Romania.

Meanwhile, the Soyuz-40 return capsule is located inside the Aviation Pavilion in the courtyard of the National Military Heritage (King Ferdinand I) Museum in Bucharest. The display features the landing apparatus with the parachute deployed. Adjacent to the display there is a photographic montage of Prunariu's historic space mission together with his space suit.

REFERENCES

1. Ionescu, Claudiu, article, *Masonic Forum* magazine, "Dumitru Dora Prunariu: A Romanian at the Heaven's Gate," issue No. 14, January 2003
2. Bert Vis interview with Dumitru Prunariu, 22nd Planetary Congress of the Association of Space Explorers, Prague, Czech Republic, 6 October 2009
3. Dumitru Prunariu email correspondence with Colin Burgess, 22 May 2014–8 June 2015

4. *Ibid*
5. Bert Vis interview with Dumitru Prunariu, 22nd Planetary Congress of the Association of Space Explorers, Prague, Czech Republic, 6 October 2009
6. Zheleznyakov, Alexander, article "Yevgeny Khrunov Remembered," *Orbit* magazine, No. 47, October 2000
7. *Ibid*
8. *Ibid*
9. *Ibid*
10. Bert Vis interview with Dumitru Prunariu, 12th Planetary Congress of the Association of Space Explorers, Montreal, Canada, 1 October 1996
11. *Soviet Weekly* newspaper, "I Saw Romania From Outer Space," issue 4 June 1981
12. Bert Vis interview with Dumitru Prunariu, 22nd Planetary Congress of the Association of Space Explorers, Prague, Czech Republic, 6 October 2009

11

Beyond Interkosmos: Soyuz T-6

With the Soyuz-40 mission, which had carried Leonid Popov and Romanian cosmonaut-researcher Dumitru Prunariu to the Salyut-6 station, the structured Interkosmos program of flying Eastern Bloc participants on week-long flights to the orbiting space laboratory was at an end. However, there were already plans to continue to fly other international 'guests' on science missions with bilateral agreements, and the participants were no longer restricted to fraternal socialist nations.

While these follow-on missions were not regarded as part of the Interkosmos program of flights, they were nevertheless seen as a practical extension to that program, and provided a means for other interested countries to participate.

As events transpired, the first of the new international space expeditions would see French '*spationaute*' Jean-Loup Chrétien visiting the Salyut-7 station in June 1982. This project was a continuation of long-standing Soviet-French cooperation in space exploration. France had been the first capitalist country to conclude an agreement with the USSR to cooperate in the study and use of outer space. That agreement was signed in 1966, during French President Charles de Gaulle's visit to the Soviet Union. In the following years this space cooperation was extended to such subjects as the nature of the space environment, cosmic meteorology, communications technologies, biology, and medicine. Joint research also included geodesy, Earth magnetism, and the study of the Sun and planets.[1]

CHILD OF THE WAR

Jean-Loup Jacques Marie Chrétien was born in the seaport city of La Rochelle in the French Republic on 20 August 1938. He was one of two sons born to Jacques and Marie-Blanche (née Coudurier) Chrétien, the other being his younger brother named Philippe. His father, a Navy sailor, would later remarry, giving the boys two younger half-brothers, Grégoire and Jérôme. Chrétien took his early education at L'École Communale à Ploujean, the College Saint-Charles à Saint-Brieuc, and finally the Lycee de Morlaix.

© Springer International Publishing Switzerland 2016
C. Burgess, B. Vis, *Interkosmos*, Springer Praxis Books, DOI 10.1007/978-3-319-24163-0_11

An interviewer from the Johnson Space Center Oral History Office once asked Chrétien about any early influences that might have led him to consider a career in aviation.

"I started many years ago, and that was during World War Two. I was a young child in Brittany, France, living a couple of miles from an airport that was occupied by the Germans, on the French shore of Brittany [and] every day I had the chance to see those airplanes, either German or British, flying over and a lot of fights. So I was living under a part of the theater of fighter airplanes of the World War Two … Of course, that probably imprinted in my mind the wish of becoming a pilot. At the end of the war, I was eight years old. I remember those last years. So I wanted to fly, to become a pilot."[2]

Jean-Loup Chrétien at the French Air Academy in 1960. (Photo: Galerie Arnaud Bard)

Chrétien began building model airplanes around the age of 14, but his concerned mother wanted to curtail her older son's enthusiasm for such a risky occupation. His beloved half-finished models would mysteriously vanish from his room. However, he was determined to become a pilot and took flying lessons at the age of 15, receiving his pilot's license a year later. He then enrolled at L'École de l'Air (the Academy of the French Air Force) at Salon de Provence in 1959, graduating two years later. That same year, 1961, he heard about the historic space flight of Yuriy Gagarin. "That's when I said, 'Okay, I want to go to space.'"[3]

In 1962 Chrétien took advanced training in the supersonic Dassault Super Mystère B-2 fighter-bomber and received his fighter pilot engineer's license later that year. By this time Chrétien had met and wed his first wife, Mary-Cathryn, and they would eventually have four children, Jean-Baptiste (1962), Olivier (1965), Emmanuel (1966) and François (1974) before the marriage ended in divorce.

Lt. Chrétien next served as a fighter pilot in the 5th Fighter Squadron, based in Orange, Provence, where he flew the Dassault Mirage III. He realized a long-held dream when he was selected in 1970 for training at his nation's test pilot school, the École du Personnel Navigant d'Essais et de Réception (EPNER) at Istres in southern France. During his seven years as a test pilot, he became chief test pilot responsible for the Mirage F-1 test program. In 1977 he was appointed as the deputy commander of the South Air Defense Division in Aix-en-Provence.

SELECTING A SPATIONAUTE

In March 1979, Soviet President Leonid Brezhnev hosted meetings in the Kremlin with his French counterpart, Valéry Giscard d'Estaing. One issue raised during talks about Soviet-French scientific and technological cooperation, was Brezhnev's offer to launch a French cosmonaut into space aboard a Soviet spacecraft. The offer was accepted and soon gained momentum. That September the French government's space agency, the Centre National d'Études Spatiales (CNES) was given the assignment of selecting two French cosmonaut-researchers to train at the Yuriy Gagarin Cosmonaut Training Center, near Moscow. An announcement was made inviting applications from suitably qualified volunteers. It was a temptation that Chrétien could not miss.

"Of course I was a candidate, and I was bothering them already, even before there were any official candidates. So I think that they knew me a little bit, and so when the official selection started, I had a little 'plus' somewhere because I had been to that administration a couple of times following summer school for space, even if I was a test pilot. So when the selection became official, I had a small advantage, I think, to all of the others. They never told me that."[4]

The selection criteria imposed by CNES stated that applicants had to:

- Be a French national
- Be aged 25 to 40 years
- Measure less than 95 centimeters and 1.81 meters sitting upright
- Weigh less than 82 kilograms
- Be in good general health.

By the closing date of 15 November, 413 applications had been received by CNES and questionnaires mailed out. The responses were mostly from airmen, seamen and engineers. Once the questionnaires and their military and medical records had been carefully perused, the number was soon reduced to 193, of which 26 were women. A screening program then reduced the list to 72 candidates. They were ordered to attend CPEMPN (Centre Principal Expertise Médicale Personnel Navigant) in Paris, the medical center of the Air Force, for extensive medical examinations. A follow-up set of tests were then carried out at LAMAS (Laboratoire de Médecine Aérospatiale) the laboratory for aerospace medicine in Brétigny-sur-Orge. After 40 candidates had been failed, the remaining 32 were returned to CPEMPN for new physical and psychological examinations. After this, only 18 candidates remained. After punishing sessions on the centrifuge, the number was further reduced to 11. Another selection phase began in December 1979, and one month later two of the remaining seven candidates were women. The next phase reduced the list to four men and one woman:

- Maj. Patrick Baudry (French Air Force), born March 1946
- Lt. Col. Jean-Loup Chrétien (French Air Force), born August 1938
- Sqn. Ldr. Jean-Pierre Joban (French Air Force), born December 1943
- Gerard Juin (Air France Airbus pilot), born April 1943
- Françoise Varnier (academic and glider pilot), born September 1949.

In February 1980, during the final stages of the selection process, the only woman still in the running, Françoise Varnier, broke her leg during a parachute accident and subsequently was eliminated on medical grounds. Varnier was killed in October 1996 during an aerobatic flight in Bourges. The four men were flown to Moscow, where they would undergo Soviet-administered tests to determine the final two candidates.

Chrétien pointed out the JSC interviewer that around that same time, he was a candidate for selection as a European Space Agency (ESA) payload specialist, with a chance to fly on future NASA space shuttle missions.

"Of course, I was a candidate, and that was for [a] Spacelab mission, and unhappily no Frenchmen were selected. That's a tough political story … the French president was very upset with that result, because France is a main contributor to ESA, and half of the French space budget goes to ESA. We pay more than double the next country, which is Germany. So, of course, that was a very strange first selection for the Spacelab mission, that no French people were selected, and the French president was very upset. So that is why I was sent to Moscow."[5]

Patrick Baudry (left) and Jean-Loup Chrétien. (Photo: Espace Patrick Baudry)

The four candidates arrived in Star City in March 1980, and during the next three months were given tuition in the Russian language and subjected to further tests. Following this they returned to France. Shortly after, on 12 June, the names of the two finalists were announced as Jean-Loup Chrétien and Patrick Baudry. The cosmonaut-researcher training would begin in Star City on 7 September.

OUT OF AFRICA

Patrick Pierre Roger Baudry was born on 6 March 1946 to Roger and Odette (née Manaud) Baudry in Douala, situated in the French colony of Cameroon in West Africa. His father was a meteorological engineer. He would grow up with two older sisters, Liliane and Nicole, and younger siblings Chantal and Philippe. After Africa, the family would settle back in France, in Eysines, Bordeaux.

Baudry took his early education at École Sainte-Marie-Lebrun à Bordeaux, then attended a military preparation school in Autun, Burgundy, where boys aged 13 to 18 were tutored for a career in the military. He then took his advanced secondary education at Prytanée Militaire de la Flèche and his college education at the Lycée Chaptal in Paris, where he was a student of advanced mathematics. In 1967, he joined the École de l'Air de Salon-de-Provence (the Academy of the French Air Force). He graduated in 1969 as an engineer and air force officer with a master's degree in aerodynamic engineering. He gained his fighter pilot's certificate and wings in 1970.

Student pilot Patrick Baudry at the École de l'Air in 1967. (Photo: Espace Patrick Baudry)

Over the next eight years Baudry served with Fighter Squadron 1/11 Roussillon, flying the F-100 Super Sabre and Jaguar jets, completing a number of missions over France and Africa. He then became a flight commander with the 50th Fighter-Bomber Wing at Toul-Rosières air base. In 1978 he trained as a test pilot at the Empire Test Pilots' School (ETPS) at Boscombe Down in Wiltshire, England. As the highest-ranking student upon graduation he received the Patuxent River Trophy, which was presented to him by Prince Charles. From 1978 to 1980 he worked as a test pilot at the Flight Test Center in Brétigny-sur-Orge, flying many different types of fighter and transport aircraft. It was while there that he was selected as a candidate for the joint Soviet-French space mission.

TRAINING REGIME

Following their arrival in Moscow, the French cosmonauts were accommodated in the same block of apartments in Star City as their Russian colleagues. Each had a three-bedroom unit.

Chrétien had actually arrived as a 'grass widow', leaving his wife and four sons in France. Baudry brought his wife Claude and their five-year-old daughter Mélodie. Claude was an architect and civil engineer, but had given up this work temporarily in order to share in her husband's experience. When asked whether she intended go back to work on their return to Paris, she replied, "I will indeed – if I can find a job."[6] Mélodie, already quite proficient in the Russian language, was to attend school and ballet lessons.

Soon after Chrétien and Baudry arrived at the Gagarin Cosmonaut Training Center, Maj. Gen. Aleksey Leonov held a press conference in which he spoke of plans for one of the two Frenchmen to fly on a joint Soviet-French space mission in 1982.

"It may seem we have lots of time," he pointed out. "Well, we haven't – if you consider the amount of work to be done by our French colleagues, that includes a number of special courses such as operating spaceship equipment and communications facilities. They must also gain basic training in space medicine, biology, and the experiments they will conduct. Then there are the constant physical exercises and workouts on simulators. And finally the Russian language. Naturally these preparations are very exacting but Zvyozdnyy Gorodok [Star City] includes all the laboratories, classes and training equipment required. All this is envisaged in the day-to-day program right up to the flight itself. The French cosmonauts will get the same training as their colleagues from the Soviet Union and other socialist countries."

Leonov was then joined by the two Frenchman, who had been attending a lecture. The journalists in attendance were eager to find out a little about their training schedule. "Our working day is simple," Baudry replied. "Up at seven, in bed at eleven, with hard work in between. Although we have got so much to do, we find time for walks and visits to other cosmonauts' families. We are received as friends here and are invited to all celebrations, both public and family. With our colleagues, we see crews off to the launch site and later welcome them back … But naturally, our work comes first."

Asked about learning the Russian language, Baudry said, "Russian is no obstacle to our studies and human contact. We feel at home in it and speak satisfactorily, so they say. We won't need any interpreters during our space flight!"

Chrétien and Baudry during a break in their training. (Photo: Author's collection)

In response to a question about which of the two Frenchmen might fly, Leonov stepped in and explained, "Which of the two men will go into space will be decided upon later. Right until the last day, our two colleagues will undergo absolutely the same training. And then it will be up to the French side to say which of the two will go – on the basis of their training results or anything else. Suppose one of them develops a cold? Things like that do crop up."

Asked about the experiments to be done aboard Salyut-7, Chrétien replied, "The French side will determine the range of space experiments and the methods. French experts will set our assignments in astronomy, medicine, biology and metallurgy. We'll be kept very busy in space too."[7]

Chrétien now says that he found the first month of training particularly hard, especially due to the separation from his wife and children, who had decided to stay at home. He was very unhappy with the regulation stating he could only return to France after two years. He and Baudry were permitted a short break during summer, but they would require to spend it in Russia. Soon after, the head of the French space agency CNES, Hubert Curien (and from 1981 the first chairman of ESA), arrived in Star City for talks. On learning of this difficult problem he spoke with administrators at the Gagarin Training Center, who finally relented and Chrétien was told that he now had permission to fly to France once every two or three months, ostensibly to carry out training on the French experiments that one of them would operate in space.

AFTER THE FIRST YEAR

Late in 1981 the two Frenchmen learned which Soviet cosmonauts they would be teamed with. Chrétien's commander was Yuriy Malyshev. Together with flight engineer Vladimir Aksyonov, Malyshev had performed the first manned flight of the new Soyuz T spacecraft with a 4-day proving flight the previous June, docking their Soyuz T-2 spacecraft with the Salyut-6 space station. The new Soyuz spacecraft could accommodate a crew of three, and Chrétien's flight engineer on the assigned Soyuz T-6 mission was Aleksandr Ivanchenkov. In 1978 he and Soyuz-29 commander Vladimir Kovalyonok had docked with Salyut-6 and established an endurance record of 139 days. Like Malyshev, that was Ivanchenkov's only space flight to that time.

At one point in their training Yuriy Malyshev was hastily removed from the Soyuz T-6 crew and reassigned to train new cosmonauts. It was later reported that the reason for his reassignment was that he did not work well with Chrétien who, being a highly skilled pilot, complained that Malyshev would not allow him to touch anything during simulations, with the result that they had exchanged angry words. On one simulation Chrétien actually took a pillow into the Soyuz simulator with him and when he was asked to explain the situation he said he might as well take a nap as he wasn't allowed to do anything. Malyshev's place as mission commander was later handed to veteran cosmonaut Vladimir Dzhanibekov.[8]

Meanwhile Patrick Baudry was teamed with commander Leonid Kizim and flight engineer Vladimir Solovyov. On his only space mission thus far, Kizim had flown to Salyut-6 aboard Soyuz T-3 in November 1980. Solovyov had yet to make his first flight into space.

A rare training photo of the original Soyuz T-6 crew of Aleksandr Ivanchenkov, Yuriy Malyshev and Jean-Loup Chrétien. Malyshev was replaced as mission commander following unresolved disagreements with Chrétien. (Photo: Author's collection)

On occasional rest breaks away from their training the two candidates found time for their respective hobbies. Chrétien enjoyed relaxing and playing music on an electric organ he had brought with him (and would later accompany him to Salyut-7). On the other hand, Baudry was adding to his collection of around two thousand labels of the best chateau wines.

The crew of Vladimir Solovyov (top of photo), Leonid Kizim and Patrick Baudry in the Soyuz simulator. (Photo: Spacefacts.de)

A year after their arrival in Star City, Lt. Gen. Vladimir Shatalov, director of cosmonaut training, spoke about the progress of the French candidates to a group of Soviet and French journalists.

"In the first year, they studied astronavigation, the dynamics of space flight, and the ship itself," Shatalov explained. "In other words, basic cosmonautics. Now we're training in the ship itself, and in January we shall begin studies in a model of the orbital station.

"Many cosmonauts from socialist countries have passed through our Training Center and we are glad that this time, too, we have quickly found a common language; have met people able to understand one another well; people who are modest and realize the diffi-culties and complexities of the joint work ahead of us. It is equally pleasant just to sit down at a table with them, laugh and joke, and do some serious work together."

Asked later in their training by Soviet journalists whether it had been difficult to master his new profession, Chrétien responded, "Space flight is a very serious matter.

Everything turned out to be much more complicated than we had supposed during the preliminary tests. Some of the highlights of my course of study were the training exercises in the Black Sea, when we donned space suits and jumped into the water; our flights in the IL-76 to simulate weightless conditions; and, of course, the various training gear on which we studied all the elements of space flight."[9]

The two Frenchmen and their respective crews were being given identical training, and neither man knew which of them would fly.

"There is no rivalry between us," Chrétien said in reply to a reporter's question. "Both of us wants to fly and our chances are equal; may the best man go up. In any case, the standby will also gain a lot, because he will get excellent training among celebrated cosmonauts. I have dreamt about this for twenty years – it is my longest dream, and now that it is coming true I can't be anything but content."[10]

Patrick Baudry was then asked for his feelings on the subject as the Franco-Soviet launch drew near.

"Of course it will not be known until the last minute which of us will go up: Jean-Loup Chrétien or myself. Given normal circumstances, he will be the first to fly. But I don't despair; I think I'll fly next time. I hope that Soviet-French cooperation in the mastery of outer space will continue, and that we shall perhaps stage more exciting experiments. It is important for both of us to carry on our cooperation. For more than fifteen years the Soviet Union and France have carried out joint work in the study of outer space. I hope that joint space flights will continue as well."[11]

With their training nearing completion the two were given a rare treat, in the form of a flying visit to Baykonur to observe the launch of Soyuz T-5 on 13 May, which was taking Anatoliy Berezovoy and flight engineer Valentin Lebedev to the new orbiting laboratory, Salyut-7, as its first residents. When asked for his impression of Baykonur, Baudry said, "The place is rather like a shrine to cosmonauts. The blast-off is amazingly beautiful and captivating. I will never forget the days spent on Baykonur.[12] In fact, that had been their second visit, because the two men had gone in April in order to inspect the Salyut-7 space station prior to its launch.

Towards the end of the training period for the two crews, an official announcement was made by the TASS news agency on the scheduled launch date for the Soyuz T-6 mission. Barring any unforeseen circumstances, it said, this would be 24 June. The announcement stated that the prime crew of Dzhanibekov, Ivanchenkov and Chrétien would join Anatoliy Berezovoy and Valentin Lebedev who had been on board Salyut-7 since 14 May, and then return to Earth on 2 July. It further stated that if the prime crew was prevented from flying for any reason, the back-up crew would receive their chance to fly. "Upon completing the training cycle at the Gagarin Cosmonaut Training Center and passing their examinations, both crews will be flown to the Baykonur Cosmodrome for pre-launch training," the TASS announcement explained.

Meanwhile, preparations continued aboard Salyut-7 with the docking of the unmanned Progress-13 on 25 May with more than 4,400 pounds of cargo, including fuel, water, food, and mail. It also transported 550 pounds of French apparatus for experiments to be carried out during the upcoming Franco-Soviet mission. Before Progress-13 was undocked for its fiery end upon re-entry, its engine was used to boost the station to a slightly higher orbit.

On 11 June the prime and back-up crews were flown to Baykonur in preparation for the launch of Soyuz T-6.

OCCUPYING SALYUT-7

On the day of the launch, the *Los Angeles Times* newspaper reported that during his flight, Chrétien, as the first French cosmonaut, would be amply provisioned with "typical French meals" as distinct from the normally-bland Soviet menu. His meals included "four kilos of specially prepared jugged hare a l'Alsacienne, crab soup, country paté, lobster pilaf with sauce a l'Armoricaine, cantal cheese, white and rye bread, candied fruits and chocolate cream. However no wines or liqueurs are included."

On a more somber note, the newspaper further reported that at the insistence of the French government, "he will fly with little or none of the political fanfare that Moscow had wanted. Owing to cool relations between France and the Soviet Union ... the nine-day mission will carry far less political symbolism than the Kremlin anticipated. The French have insisted that the Soviets treat the mission as a purely scientific exercise."[13] This came about because relations between the two nations had cooled significantly since the former French President, Valéry Giscard d'Estaing had helped to organize the joint mission back in 1979, shortly prior to the Soviet invasion of Afghanistan. His successor, François Mitterand, who took over the presidency in May 1981, decided to let the flight go ahead on the proviso that it was without all the political pomp of the earlier Interkosmos flights.

In compliance with Mitterand's directive, no French minister would attend the launch and Chrétien was instructed not to participate in the traditional propaganda salutes to peace and progress that traditionally occurred prior to the launch. As one French official explained it, "There will be no medals, no music, no hugging and kissing. We're trying to minimize the symbolism."[14]

As the rocket left the pad, shown live on Soviet and French television, Chrétien seemed to have momentarily forgotten all the Russian he had learned, shouting out in joy the traditional French fighter pilot's rallying cry "A la chasse!" which translates as "The hunt is on." Soon after though, he appeared to have some anxious moments. Cameras on board showed him throwing worried glances at his companions. Vladimir Dzhanibekov reported "some slight vibrations" a few seconds later, but then said everything was fine.

The crew docked with Salyut-7 the next day, and once the transfer hatch had been opened, the three men floated into the space station, where they were welcomed by the two residents, Berezovoy and Lebedev.

After a brief rest period, the visitors began their joint research program, which included a range of experiments in space medicine and biology, technology tests and physico-technical experiments.

Over the next few days, all five crewmembers successfully carried out a comprehensive program prepared jointly by Soviet and French scientists – the latter involving over twenty French research institutes and laboratories. A large portion of this program was devoted to research into the effects of weightlessness on the human organism. This involved constant observation and accumulation of data for a detailed study of how the cardiovascular system undergoes certain changes as the body adapts to the space environment.

The Echography experiment used a French ultrasonic apparatus for a study of the heart's functions, the velocity of the blood flow through major blood vessels, and changes in the size of the latter during adaptation to weightlessness. In the Posture experiments the cosmonauts also studied the interaction between human sense organs and motor system. For the Neptune experiment they checked the sharpness and depth of their vision, the

Lift-off for the Soyuz T-6 mission to Salyut-7. (Photo: Spacefacts.de)

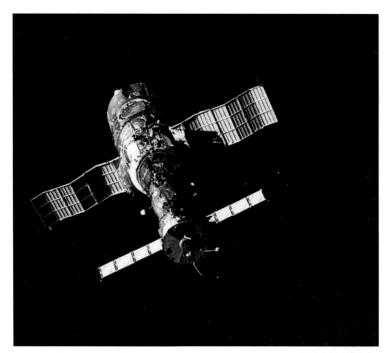

The Salyut-7 space station as seen from Soyuz T-6. (Photo: Spacefacts.de)

sensitivity of the vision analyzer (the Mars-2 experiment), and hygienic conditions on board Salyut-7 with two crew members present in the Microbe Exchange experiment. The cosmonauts also used a special prophylactic device, the Bracelet, to study the possibility of normalizing blood circulation in conditions of weightlessness.

Of particular interest to scientists on the ground were results obtained in the Biobloc-3 and Cytos-2 experiments. The purpose of the Biobloc-3 was the study the effect of heavy cosmic particles on biological objects. The data obtained would extend contemporary knowledge of the effects of cosmic radiation, provide a measure to assess the danger of cosmic irradiation for biological systems, and assist in developing appropriate protection, which is particularly important on long-duration space missions. The Cytos-2 experiment studied the antibiotics sensitivity of bacteria cultivated in vitro during the flight. It was hoped the results would be able to be used in the development of methods for the prevention and treatment of different infectious diseases that might occur during space flight.

The crew also carried out three vital technological experiments – Gauge, Diffusion, and Liquation – with the aid of a Soviet Kristall furnace. The Gauge experiment consisted of checking the temperature field of an electric oven under various modes of operation, and the simultaneous registering of micro-acceleration forces along the station's axis. The Diffusion experiment was aimed at determining the diffusion coefficients of two materials (in this case copper and lead) in weightlessness. The Liquation experiment studied the effect of capillary forces on the formation of an alloy of aluminum-indium – metals which

Chrétien in a playful mood aboard Salyut-7. (Photo: Spacefacts.de)

do not blend under normal terrestrial conditions. This experiment was also of practical value for the production of new composite materials made of components having considerably different densities and melting temperatures. Trying to do this on Earth is extremely complicated because gravity causes the two materials to separate.

In addition, Chrétien and other crewmembers took photographs of different regions of the Earth's surface and oceans in accordance with a program aimed at studying natural resources and the environment.[15]

By 2 June, Dzhanibekov, Ivanchenkov and Chrétien had completed their work aboard Salyut-7 and had prepared their spacecraft for the return journey.

Meal time aboard the space station. (Photo: Author's collection)

BACK HOME

The Soyuz T-6 descent module carrying the three cosmonauts returned safely to Earth at 6:21 p.m. Moscow time on 2 July, landing some 40 miles north-east of Arkalyk. The cosmonauts were reported to be "in good health" by medical personnel in the Soviet recovery team.

The three men were flown to Star City for a full medical check-up. This established that they were suffering no ill-effects from their eight days in space. At a later press conference, Chrétien took the opportunity to say, "Soviet technology functioned admirably."

The Soyuz T-6 descent module making a successful landing. (Photo: Spacefacts.de)

French President Mitterand spoke highly of the results of the joint mission, saying it was testimony of over fifteen years of developing fruitful cooperation between the two countries in the study of outer space for peaceful purposes, and demonstrated a constant aspiration for mutual benefit. "It is a symbol of the French people's goodwill to help in building peace in Europe on the basis of traditional friendly relations and respect for the rights provided for in the Helsinki agreements, and to help develop good relations among all European nations."

Scientists and specialists regarded the joint mission as the beginning of a new stage in Soviet-French cooperation in science and technology.

"Scientists from our two countries jointly elaborated sixteen research technical projects," explained Vyacheslav Balebanov, deputy director of the Institute of Space Research of the USSR Academy of Sciences in response to a reporter's question. "The cosmonauts staged astronomical and geophysical experiments in orbit, carried out research into space biology and medicine, and also conducted more immediate, applicable experiments. For instance, in conditions of weightlessness, new alloys and medicines are obtainable, more effective than can be made on Earth."[16]

"The research is of great importance," added Nicole Moati, a French specialist. "And for terrestrial medicine as well. The data obtained can be used in all cases when man has to live for long periods in conditions of closed space; for example, on polar expeditions."[17]

The Presidium of the USSR Supreme Soviet awarded Vladimir Dzhanibekov the Order of Lenin and Aleksandr Ivanchenkov the Order of Lenin and a second Gold Star medal. Jean-Loup Chrétien was awarded the title of Hero of the Soviet Union, the Order of Lenin and the Gold Star medal.

After Baudry returned to France, CNES officials said he had performed exceptionally well and would be assigned to the next space flight opportunity. Meanwhile he continued to fly in the French Air Force. In that role he came perilously close to losing his life on 4 June 1983. That day he took off from the French airport at Morlaix with passengers Jean-Loup Chrétien and invited guest cosmonaut Anatoliy Berezovoy aboard. Soon after take-off the light plane that he was flying suffered mechanical problems and crash-landed in a field of cauliflowers. Fortunately all three men escaped with minor injuries.

FRENCHMAN ON A SPACE SHUTTLE

In April 1984 Patrick Baudry was chosen to become the first French *spationaute* to fly on an American space shuttle. That October he traveled to the United States to undertake a year of NASA training for the STS 51-E mission. Jean-Loup Chrétien served as his back-up. With just six days remaining to the scheduled launch, the mission was canceled owing to technical problems. The crew was then reassigned to mission STS 51-D with the exception of Baudry, who transferred to STS 51-G, which launched on 17 June 1985 aboard shuttle *Discovery* on a seven-day mission.

Baudry and Chrétien training as NASA payload specialists. (Photo: Spacefacts.de)

The international crew aboard *Discovery* deployed communications satellites for Mexico, the Arab League, and the United States, and then deployed and later retrieved the SPARTAN satellite which performed 17 hours' of observations for X-ray astronomy. Baudry's principal responsibility was a scientific and medical program for both French and American research laboratories, in particular the Sonography experiment, a direct study of the adaptation of the cardiac muscle to weightlessness. Another experiment investigated the eyes, inner ear, and muscles as the vestibular (balance) system adapts to weightlessness.

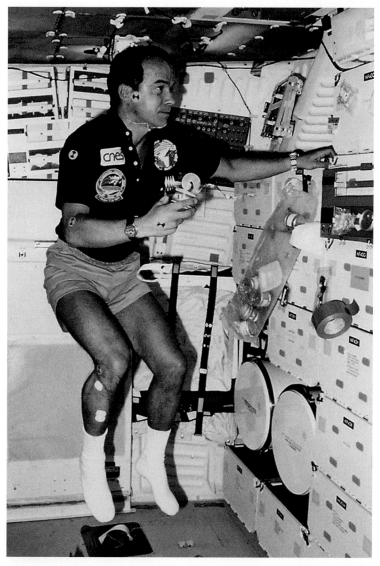

Baudry enjoying a mid-deck snack aboard shuttle *Discovery*. (Photo: NASA)

Discovery landed at Edwards Air Force Base in California on 24 June. In completing this flight, Baudry traveled 2.5 million miles in 112 circuits of the Earth, logging over 169 hours in space. However it would prove to be his one and only space flight.

On returning to France in 1985, Baudry resigned from the CNES astronaut team and took employment with the Flight Research Center, CEV (Centre d'Essais en Vol), in the town of Brétigny-sur-Orge. In 1986 he was appointed Advisor to the Chairman of the

Comparative photos showing Baudry and Chrétien at the time of their selection as CNES cosmonauts, and together again at the 80th anniversary celebrations for the French École de l'Air on 7 May 2014. (Photos: Author's collection and Patrouille de France)

Aerospatiale Company. He was also appointed chief test pilot for the European space plane Hermes, for which Aerospatiale was the industrial prime contractor. In 1989 he designed and created the European Space Camp for young people. It had to close in 1992 owing to a lack of political support but by then Space Camp Patrick Baudry de Cannes Mandelieu had hosted more than 10,000 young people. From 1993 to 2003 he served as an advisor for manned space flight at EADS (European Aeronautical Defense Systems), and is now a test pilot at Airbus Industrie and president of the school group ACADIS. This is the French acronym for a special school in Bordeaux that aims to help both struggling and talented students who can take scholastic advantage of small classes combined with targeted teaching. He also spent time working in several African countries for UNESCO (United Nations Educational, Scientific and Cultural Organization), and on 6 September 1999 in appreciation of his many years of work with the organization, the Director-General of UNESCO, Federico Mayor, awarded Patrick Baudry the title of 'Ambassador of Goodwill'.

When Patrick and Claude eventually ended their marriage, this prompted Baudry to make the observation, "Going into space is easier than getting divorced." He is now married to his second wife, Stéphanie. A prolific writer, he has several published books to his credit on the subject of space exploration.

Among his numerous other recognitions, Baudry is a board member of the International Olympic Truce Foundation, and he has been presented with the Chevalier of the Legion of Honour; the Chevalier of the National Merit Order; the French Astronautics Medal; the Soviet Order of Friendship of Peoples, and the Soviet Order of Gagarin. He was also awarded the American Space Flight Medal and Lindberg Trophy (USA).

As related in Chapter 16 of this book, Jean-Loup would fly into space once again in 1988 as cosmonaut-researcher aboard the Soyuz TM-7 mission to the Mir space station.

REFERENCES

1. Gennadiy Ryabov, *Soviet Weekly* article, "New International Space Mission," issue 2 July 1982
2. Jean-Loup Chrétien interview with Carol Butler for NASA JSC Oral History Program, Houston, Texas, 2 May 2002
3. *Ibid*
4. *Ibid*
5. *Ibid*
6. Svetlana Soldatenkova, *Moscow News* article, "No Rivalry – May the Best Man Go Up," issue 27 May 1981
7. I. Nekhamkin, "Newcomers to Zvyozdnyy," *Soviet Union* magazine, issue July 1980
8. Tim Furniss, David J. Shayler and Michael D. Shayler, *Praxis Manned Spaceflight Log, 1961-2006*, Springer-Praxis Publication, Chichester, U.K., 2007, pg. 257
9. Gennadiy Ryabov, *Soviet Weekly* article, "New International Space Mission," issue 2 July 1982
10. Svetlana Soldatenkova, *Moscow News* article, "No Rivalry – May the Best Man Go Up," issue 27 May 1981

11. Tatyana Butkovskaya, *Moscow News* article, "Our Space Crew is a Single Entity," issue 22, 6 June 1982
12. *Ibid*
13. *Los Angeles Times* unaccredited article, "First French astronaut will travel a la carte, issue 24 June
14. Marc Frons and Timothy Nater, *Newsweek* magazine article, "Space-Age Politics with a Gallic Touch," issue 5 July 1982 (Vol. 120, No. 1)
15. Gennadiy Ryabov, *Moscow News* article, "Mission Concluded," issue 9 June 1982
16. Svetlana Soldatenkova, *Moscow News* article, "Soviet Technology Functioned Admirably," issue 9 June 1982
17. *Ibid*

12

Soviet-Indian mission

It was towards the end of the tumultuous sixties that India began drawing up its national space program with the assistance of the Soviet Union in the form of both technical advice and equipment. Specifically, Soviet scientists and technicians helped India to develop the designs, service systems and research facilities for its satellites, as well as to construct the necessary space centers and ground control complexes.

The first indigenously built Indian satellite, the *Aryabhata*, lifted off from the Kapustin Yar Cosmodrome in the Soviet Union on 19 April 1975 atop a Kosmos-3M launch vehicle and was placed into a low-Earth orbit. Developed and built by the Indian Space Research Organization (ISRO), it was later followed by the *Bhaskara* and *Bhaskara-2* remote sensing satellites; also built with Soviet assistance and, like their predecessor, placed into orbit by a Soviet rocket. In 1982 an agreement was reached which would see a series of large Indian satellites placed into solar-synchronous orbits to enable ISRO to continue its exploration of the Indian subcontinent's natural resources from space. The first of these satellites, IRS-1A, was launched on 17 March 1988 aboard a Vostok-2M rocket.

A JOINT FLIGHT IS PROPOSED

The concept of flying an Indian cosmonaut aboard a Soviet spacecraft had been first mooted in June 1979, when the Soviet Union offered India the opportunity to fly a cosmonaut on a joint mission. This proposal was confirmed in May 1980 during an announcement by Yuliy Vorontsov, the Soviet Ambassador to India, disclosing that the two countries had signed off on a massive new arms deal worth around $US1.6 billion. As reported by Bruce Loudon of the *Daily Telegraph* in the U.K., this also coincided with the arrival in New Delhi of the first installment of 'heavy water' from the USSR, for use at the Rajasthan atomic power plant in Kota. The ambassador insisted that the launching of an Indian cosmonaut was for "peaceful research" alone and had nothing to do with military cooperation.

The prospect of a joint Indian-Soviet flight was once again brought to the fore by Soviet President Leonid Brezhnev during his official visit to India in December 1980. The offer was formally accepted with gratitude in March 1981, with the support of India's Prime Minister Indira Gandhi. Speaking in Parliament, she said that India could not remain aloof from the development of science, which is why the Indian people should appreciate the

© Springer International Publishing Switzerland 2016

C. Burgess, B. Vis, *Interkosmos*, Springer Praxis Books, DOI 10.1007/978-3-319-24163-0_12

Indian Prime Minister Indira Gandhi with Soviet President Leonid Brezhnev. (Photo: Source unknown)

opportunity rendered by the Soviet Union because it would enable Indian scientists to widen the scope of their knowledge. A program was then set in place to select a suitable pair of candidates who would train for the mission in Moscow.

The Indian Air Force was asked to choose two cosmonaut candidates possessing test pilot experience, and in 1984 the choice narrowed to Ravish Malhotra, 40, and Rakesh Sharma, 36, both of whom held the rank of wing commander.

THE SOUND OF JETS IN THE SKY

Rakesh Sharma, the son of Tripta and Devendranath Sharma, was born into a Punjabi family on 13 January 1949 in the district of Patiala in the Republic of India. As a young boy he was raised in Hyderabad and enrolled at St. George's Grammar School.

As far back as he can remember, Rakesh (with the nickname Rikki) wanted to be a pilot. He recalled that when he was six years old the Indian Air Force received its first jet fighter, the de Havilland Vampire, with its unusual twin fuselage structures. "I used to hear it fly overhead," he reflected of that time. The Air Force training establishment was in Hakimpet, and the Vampire was based there. "I would cycle some ten to fifteen kilometers from my home to the base, stand outside the fence, and watch it fly; I was pretty hooked to this."[1]

Rakesh had a cousin in the Air Force who gave him a tour of the air base, showing him different aircraft at close quarters. He was also permitted to sit in the cockpit of a Vampire jet fighter. At the age of seven, he was certain that he wanted to be a pilot. "If you end up doing what you are passionate about, the journey is so easy," he concluded.[2]

A determined student, Sharma was always in the top five of his class in Nizam College, also in Hyderabad. Away from his scholastic endeavors he was fond of the outdoors and quite proficient at cricket, making it into the all-Indian inter-university team.

As a fifteen-year-old undergraduate student, he qualified for the National Defense Academy (NDA) in Khadakwasla, where he enrolled in 1966 as an air force cadet, having opted to serve in the Indian Air Force. He would spend the next four-and-a-half years at the academy. While he was there he met his future wife, Madhu. Following some professional training and gliding he then began flight training. On 13 January 1970 Sharma was commissioned an officer in the Indian Air Force, and was married soon after. He then served with various squadrons, progressing through the ranks.

Rakesh Sharma. (Photo: Spacefacts.de)

Sharma was only 23 years old when the Indo-Pakistan war broke out in 1971. His first operational aircraft was the MiG-21, and he flew 21 air defense and interception missions. He was then selected for test pilot training, which he completed. Then he went to work for the Aircraft and Systems Design Establishment in Bangalore, where he was operating such aircraft as the English Electric Canberra, the Hawker Siddeley HS-748, the Hawker Hunter, the de Havilland Caribou, the Polish Iskra jet trainer, and Indian Hindustan Aeronautics Ltd. (HAL) aircraft such as the HPT-32 Deepak, Kiran, Ajeet, and Marut. During this period he managed to log over 1,600 hours of flight time.

FIGHTER PILOT AMBITIONS

Ravish Malhotra was born on 25 December 1943 in the Punjabi city of Lahore, once part of British India, but nowadays a major city in Pakistan. After the turbulent partition of August 1947 in which the Islamic Republic of Pakistan was founded in the north and the Republic of India was established in the south, the Malhotras moved into Calcutta (now Kolkata), where Ravish attended the St. Thomas High School. After graduating in 1959, he enrolled as an air force cadet at India's National Defense Academy in Khadakvasla. "I wanted to be a fighter pilot," he pointed out in a rare interview in 2005. "But then, I didn't realize I would end up on a space program."[3]

Ravish Malhotra. (Photo: Spacefacts.de)

After his graduation from the defense academy in 1963, Malhotra was commissioned in the Indian Air Force and became the fighter pilot he had dreamed of as a child. He then took an active part in the 1965 and 1971 hostilities with Pakistan, serving in a number of different operational squadrons. In 1970 he qualified as a flight instructor. During the 1971 conflict he flew eighteen operational missions in Su-7 ground-attack aircraft with No. 26 Squadron, based at Adampur. One of his more dangerous missions, in which he came close to being shot out of the sky, was described by former squadron member Flt. Lt. Narayanan Menon:

> On 16th December, [Wing Cdr.] R. K. Batra, the [commander of the squadron], led a four-aircraft bombing mission to the railway yard at Narowal. Flt. Lt. T. S. Dandass, a classmate of mine from Delhi, was No. 2; Flt. Lt. Ravish Malhotra, who later trained as a cosmonaut, was No. 3; and I was No. 4. The mission was flown late in the afternoon, and visibility conditions were poor. The aircraft were heavily loaded and it was to be a steep glide attack from 4.2 kilometers height to give greater accuracy as the intent was to destroy the Narowal rail yard, a prime interdiction target. Due to [a] navigational error, the formation drifted left of track and when we pulled up the target was displaced well to our right. The speeds had dropped during the pull-up and we were lazily turning right to get into attack mode when the sky was filled by anti-aircraft fire. Despite our altitude, the ack-ack shells were bursting around and above us. The barrage of fire came from Chinese-built quads. We rolled into the attack through the ack-ack fire and went into [a] steep dive individually. Being the last, I saw No. 1 pulling out after bomb release and No. 2 in the dive. A few seconds later No. 2's aircraft appeared to continue in the dive and then impacted with the ground. I saw the huge ball of fire followed by a plume of smoke. By now I had achieved release conditions. I released my bombs and pulled out of the dive. Later, when the formation was gathering up, the leader asked for check-in on the radio. All except No. 2 checked in but the time was not right for me to volunteer any information because the images were still fuzzy in my mind. On landing, I related what had happened. I could not confirm an ejection as I did not see any. Later it was conjectured that the aircraft could have taken a direct hit on the cockpit, disabling the pilot.[4]

Around this time Malhotra married; he and his hotel manager wife Mira (who is now a psychologist) would later have two children; a daughter Rakhi, and a son, Rohid.

In 1974 Malhotra went to the United States to attend the prestigious U.S. Air Force Test Pilot School (TPS) at Edwards Air Force Base in California, joining the 24-strong Class 74A. Two of his fellow trainees, John Casper and Ron Grabe, were subsequently hired by NASA as astronauts; both achieved the rank of colonel and both commanded space shuttle flights, each with four missions to his credit.

After returning to India as a qualified test pilot, Malhotra became a commanding officer stationed at the test pilot school in Bangalore and was also the Air Officer Commanding of Hindon Air Force Station near Delhi.

A serving major in the Indian Air Force at the time of the Soviet-Indian selection process, Malhotra was one of 240 applicants. Among other requirements was that a selectee had to be a test pilot, and he had not only graduated from the U.S. Air Force Test Pilot School, but was also the chief test pilot with the Indian Air Force Flight Test Organization and by this time he had test-flown the Vampire, MiG-21, Ajeet, Canberra and Kiran aircraft, and logged around 3,400 hours of flying time.

COSMONAUT TRAINING CENTER

The two selected candidates would undertake their training at the Yuriy Gagarin Cosmonaut Training Center near Moscow. "I went to Moscow in September 1982 with my wife Madhu and two children," Sharma reflected. Their training course began on 23 September. "It was cold but the cosmonauts and their families were warm and friendly."[5]

The two Indian families experience wintery conditions in Moscow. (Photo: Author's collection)

Malhotra (left) and Sharma inside a Soyuz training module. (Photo: Author's collection)

As Boris Volynov, a former cosmonaut and senior administrator at the cosmonaut training center stated at the time of the Indian trainees, "They are goal-oriented people. Ravish and Rakesh came to us with no knowledge of Russian. Within a short time they not only learned it, but developed a good command of it; they take their lecture notes, read documents and take exams – all in Russian."[6]

"Our working day begins at nine a.m. and ends at six fifteen p.m.," Sharma added. "We get up early and go to bed late. Now we play tennis in our off hours, and in the winter we made ski outings in the outskirts of the Stellar Township [Star City]. This was our first real Russian winter – with frosts."[7]

Madhu Sharma (her first name meaning 'honey') was by profession an interior design artist; their son Kapil (then eight) and daughter Mansi went to school in Moscow. Sadly, Mansi Sharma died following what was termed "post-operational complications" from an undisclosed illness on 11 May 1983, aged just six. It was a measure of his dedication that, while overcoming his grief, Sharma continued with his cosmonaut training.

Altogether, the training for the Indo-Soviet space flight would take over 18 months, and the preparation of Sharma and Malhotra was divided into two phases. First they had to learn the Russian language, acquaint themselves with the fundamentals of space navigation, bio-medicine, and biology, Earth resources, and materials sciences, and familiarize themselves with the stellar sky and spacecraft control systems.

"Every minute is counted, and classes are conducted like in an institute," said Volynov. "There are no exceptions or privileges for our guests. The Center already has more than twenty years' experience in work with foreign colleagues. For after preparation in Star City they don't go to outer space as passengers but perform definite duties aboard the spacecraft like other members of the crew."[8]

The rigorous training included flights on a converted IL-76 airplane, simulating conditions of weightlessness by flying a series of dizzying parabolic arcs. They also used simulators to rehearse procedures for docking and undocking and re-entry. There were important survival techniques to practice in case they had to return to Earth in an unplanned location, and they improved their fitness through physical conditioning and sports exercises. Throughout their training, and as part of the conditioning process, almost everything in their lives was closely monitored.

"The Russians thought it would be tough for us," Malhotra stated, "but I thought it was pretty normal. So while we had a 7:00 a.m. to 7:30 p.m. schedule, we couldn't eat dinner at home because even our calories had to be calculated. That way, once in space, no one could say they were feeling sick and wanted to get back home."[9]

As Volynov noted, all of the training and lessons were conducted in Russian. Although they found learning the language difficult, both trainees gained adequate proficiency as the months went by, as Sharma revealed. "My wife and I learnt enough Russian to venture out on our own for sightseeing and shopping expeditions. [We returned home with] delightful memories."[10]

Maj. Gen. Andriyan Nikolayev, deputy chief of the training center, was evidently pleased with their progress. "Both candidate cosmonauts have passed the exams on the fundamentals of space navigation and space medicine with flying colors," he said in one interview. "They did well in the course on the study of the starry sky and on spaceship control systems."[11]

Indian cosmonaut-researcher candidates Ravish Malhotra and Rakesh Sharma. (Photos: Spacefacts.de)

TRAINING INTENSIFIES

In June 1983 Sharma and Malhotra underwent splashdown training in Fedosiya, on the coast of the Black Sea. After a short break the following month in India, they returned to Star City and were teamed with their respective commanders and flight engineers. Sharma was teamed up with Col. Yuriy Malyshev and flight engineer Nikolay Rukavishnikov, and Malhotra got Anatoliy Berezovoy and Georgiy Grechko. Thereafter they began the second phase of their preparations with all members of each crew working together in order to master the systems of the Soyuz T spacecraft.

Splashdown recovery training. Photo shows Nikolay Rukavishnikov (far left) who would later be replaced in the prime crew due to illness. (Photo: Author's collection)

On 6 October 1983 it was announced that the prime crew for the Soyuz T-11 mission was the one commanded by Malyshev. It was felt that Sharma was selected as the more suitable candidate of the two because he had demonstrated a better command of the Russian language than Malhotra. Later in their training, Rukavishnikov failed his medical due to what was said to be a bad case of the flu. He was replaced by Gennadiy Strekalov. Meanwhile the mission emblem selected for the Soviet-Indian mission was the Sun God, Surya, depicted in a chariot being drawn across the sky by horses.

One of the experiments began three months prior to the scheduled launch, when Sharma and Malhotra dropped the rigid fitness routine which the Soviet cosmonauts were following, and instead prepared an elaborate series of intensive yoga exercises to assess whether yoga might assist other space travelers to cope better with weightlessness, and even overcome the insidious and largely unknown malady of 'space sickness' that impaired the performance of some astronauts and cosmonauts during their first few days in space.

Malhotra and Sharma performing exercises with their yoga instructor. (Photo: Author's collection)

One of the major investigations to be carried out during the upcoming mission was known as the Terra experiment. According to Aleksandr Koval, a department head of the national Priroda (Nature) studies center, "The program of the Indo-Soviet flight includes in particular the taking of a series of pictures of India's territory. The objective of the Terra experiment is to study the natural resources of the subcontinent and the adjacent areas of the Indian Ocean. With an area of 1.23 million square miles, India has a rather varied geography and climate. Indian experts are particularly interested in the Himalayas. This area has not been studied well enough, although many of India's rivers begin right there. India has plans to increase production of electricity by building alpine power stations. Indian experts would like to use Soviet experience in building such stations. Clearly, one ought to have as much information about water resources in order to build future power stations.

"Information about water resources is also important to develop farming. Practically all arable land in India is already being used. Its productivity can only be increased by finding more water. This is important both for agriculture and animal husbandry. With that in mind, Indian experts requested to pay particular attention to the Thar Desert when taking pictures from outer space. There is a search for subterranean sources of water and an investigation of the so-called seasonal watercourses.

"India is lacking in such important natural resources as oil and gas, therefore the Terra experiment is also designed to try and find promising geological structures in this respect. India's forestry experts will be greatly interested in the pictures from space, because they would like more up-to-date information on the timber reserves, the location of forests and possible routes of timber haulage.

"As regards the Indian Ocean, the focus, naturally, will be on finding areas that are rich in plankton. On the whole, the Terra program covers all major aspects of nature studies."[12]

The back-up crew in training: Grechko, Berezovoy and Malhotra. (Photo: Spacefacts.de)

The day they left Star City bound for the Baykonur launch complex in Kazakhstan, the cosmonauts made their way to Yuriy Gagarin's study and signed a diary; a ritual still carried out before every manned Russian mission. Sharma and Malhotra went through another Indian ritual as their wives put tilak marks on the men's foreheads for good fortune, before waving them on their journey.

For ten days prior to the scheduled launch date, Sharma and Malhotra were kept isolated in germ-free enclosures. The day before the launch, the names of the prime crew who would fly the mission were formally announced and remained unchanged. Rakesh Sharma would become the first Indian citizen to fly in space.

The prime and back-up crews relax prior to the mission. (Photo: Author's collection)

ACTIVITY ABOARD SALYUT-7

On 3 April 1984, a freezing spring morning at the Baykonur Cosmodrome on the vast grayish-brown steppes of Soviet Kazakhstan, the Soyuz T-11 spacecraft was perched atop a 14-storey rocket with its crew aboard. Two-and-a-half hours earlier, the three cosmonauts had emerged from a germ-free enclosure in their space suits before boarding a bus for the two-mile drive to the launch pad. As the distant crowd at the site cheered and waved, Yuriy Malyshev formally announced, "We are prepared to complete the mission." Once they were given permission to fly, they rode the elevator to the top of the rocket where they were inserted into the spacecraft and sealed inside.

As Yuriy Malyshev, Gennadiy Strekalov and Rakesh Sharma waited for lift-off, millions of people in India were watching live images on their television sets. At precisely 6:38 p.m. (Indian Standard Time) their patience was rewarded when a computer in the control center sent a small electrical pulse that ignited the 300-tonne rocket's liquid oxygen and kerosene propellants.

A muffled roar quickly spread across the cosmodrome as a thrust equivalent to two million horses lifted the rocket off its support structure with what seemed agonizing slowness. As the engines built up more thrust, the slender rocket rose steadily into the sky, finally disappearing from sight, leaving behind a trail of rapidly vanishing white smoke.

In the control center the launch and ascent were viewed on a computer console by back-up Ravish Malhotra and members of the Indian delegation. Madhu Sharma, her nine-year-old son Kapil, and Mira Malhotra witnessed the event on color television sets in their apartment complex at the Star City training center.

With the launch under way, India now became the 14th nation to send a person into space. Sharma, the 138th space traveler, carried with him a small amount of soil from Raj Ghat (the place where Mahatma Gandhi was cremated), portraits of Gandhi, Pandit Jawaharlal Nehru, President Zail Singh, Prime Minister Indira Gandhi, and Ramaswamy Venkataraman who at

The crew waves before boarding a transfer bus out to the Baykonur launch pad. (Photo: Author's collection)

that time was the Defense Minister (and would later be the nation's President). He was also carrying some Hindi films and recordings of sitar music, as well as fresh mangoes and other Indian foods for all the crew to consume aboard the Salyut-7 station.

Ravish Malhotra was later asked if he was disappointed not to have been chosen to make the flight. "Whether it is Sharma or me, it is an Indian who is flying," he very diplomatically observed. "I'm elated."

Twenty-five hours after lift-off, the Soyuz T-11 spacecraft successfully docked with the orbiting Salyut-6/Soyuz T-10 space complex. Once aboard the station they were greeted by Leonid Kizim, Vladimir Solovyov and Oleg Atkov, who had taken up residence more than two months earlier. Once all of their gear had been transferred and stowed, the newcomers joined the residents for a celebratory dinner. Then it was time to knuckle down to the work which was to occupy them for the next seven days.

Said Sharma, "There was so much hectic activity on board the spaceship, so many things that each of us had to do, that we literally had no time to sit around and stare into space."

The mission was both busy and varied. As well as environmental studies, photography, and investigations into materials technology, it put great emphasis on medical experiments with the aim of learning more about the effects of microgravity on the human organism. It is hardest for space travelers during the so-called period of acute adaptation to weightlessness, in the first week of a flight when an abnormal flow of blood to their heads and lungs causes vestibular and circulatory disorders and leads to a marked decrease in overall efficiency. A series of experiments was carried out to establish what happens to the cardiovascular system. The Ballisto experiment, for instance, involved recording the

Sharma and Malyshev (top) with resident Salyut-7 crewmember Vladimir Solovyov (left) and Oleg Atkov. (Photo: Spacefact.de)

micro-movements of the body which are part of cardiac activity. The experiment yielded data on the magnitude and spatial distribution of systolic energy, and on how conditions during a space mission influence this distribution. Such data were of great scientific value because they provided a deeper insight into the processes of vibratory energy conversion along three axes and contained diagnostic information on the state of the systolic function of the heart, of its right and left ventricles. Another experiment, Vector, was to study the bioelectrical activity of the heart.

A yoga experiment was conducted in orbit for the first time. Sharma had studied diverse yoga positions or 'asanas' before the flight. Five postures were selected for the experiment and he carried out these exercises every day in space. It wasn't an easy thing to accomplish in weightlessness; he had to secure himself into position by finding a rigid point of support. Sharma strapped himself to the space station's gym apparatus. The experiment provided a wealth of information on the activity of the back, hip and shin muscles in free exercises, on the biomechanics of various groups of muscles involved in performing the asanas, and also on the specific features of muscle control and coordination in space flight conditions. The motor and bio-electrical activity of the muscles was periodically assessed by using special recorders.

"It is interesting to compare various systems of physical training," Air Marshal Mulk Raj, Chief of the Indian Air Force Medical Service, told reporters in the control center. "We are of the opinion that yoga exercises can reduce the cardiovascular disorders which are caused by weightlessness. In addition, these exercises may prove an effective means of preventing the muscular atrophy which develops in the state of weightlessness during long flights."[13]

While working aboard the Salyut station, Sharma enjoyed a televised conversation with Prime Minister Indira Gandhi via a satellite video link-up. At one stage the Prime Minister asked how India looked from in space. Sharma then immortalized himself in many Indian hearts when he replied, "Saare Jahan Se Achha," which translates as "the best in the world."

RESEARCH CONTINUES

The onboard work continued almost unabated. The information from the Terra program of research into India's environment and natural resources was of particular importance for the nation's economy. Following the instructions of geologists, soil scientists, and farming and water management experts, the cosmonauts observed and photographed the subcontinent and the adjacent ocean. They photographed the Nicobar and Andaman islands with a view to detecting shoals that might yield oil and gas, inspected the forested areas and tree plantations in the central part of the subcontinent, the Ganges River basin, the glacial and snow cover of the Himalayas, and individual ocean areas in order to determine their biological productivity.

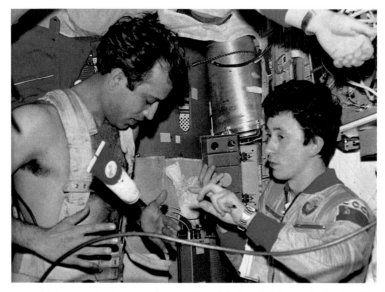

Sharma and Atkov conducting experiments aboard Salyut-7. (Photo: Author's collection)

Thousands of photographs were taken from space and, simultaneously, from a laboratory aircraft. This was backed up by land and offshore investigations of regions with formations that were representative of the rest of the subcontinent. The information gathered during this experiment would be used to draw up land usage and coastal zone control maps, as well as in cartography, oceanographic research, the study of the state of forests, inland water bodies and sown areas, mineral prospecting, the building of electric power stations, roads and irrigation canals.

Additionally, the crew carried out several materials science experiments designed to study the phenomenon of 'overcooling' during the solidification of molten metals. This explored the possibility of exploiting conditions of micro-gravity to produce special forms of metallic materials known as 'metal glasses' which are impervious to radiation, high temperatures and aggressive media. The source material was a silver-germanium alloy prepared by an Indian team. Scientists hoped that in space the production of glass-like metals would require only heating and coating. If the results were encouraging, they felt it might be possible to set up commercial production of these valuable materials in space in order to facilitate progress in many spheres of science and engineering.[14] By way of an explanation, in a metal the atoms are generally arranged in a crystalline lattice, whereas a glass, even if chemically similar, is an amorphous (non-crystalline) solid. A glass is formed by cooling the melt so rapidly that crystals cannot form.

END OF A MISSION

The Soviet-Indian space mission came to an end on 11 April when Malyshev, Strekalov and Sharma returned to Earth in Soyuz T-10, leaving their own spacecraft at the Salyut station. The six-man international team had successfully worked together in the orbital complex for seven days. Clearly delighted with the success of the mission, Indira Gandhi remarked that the flight was an important event in the history of relations between the two nations.

The three cosmonauts are all smiles after landing. (Photo: Author's collection)

Post-flight, Rakesh Sharma was conferred with the honor of Hero of the Soviet Union, as well as the Ashok Chakra, an Indian military decoration that had replaced the British George Cross and is awarded for valor, courageous action or self-sacrifice away from the battlefield.

The Indian people gave an enthusiastic welcome to the nation's first cosmonaut, his back-up Ravish Malhotra, and their Soviet associates led by Lt. Gen. Vladimir Shatalov, in charge of cosmonaut training, as they began a triumphant ten-day tour of the country. Adulation and the mandatory round of functions and speeches particularly followed Sharma's return home, but the assassination of Indira Gandhi on 31 October 1984 in New Delhi put an immediate and massive dampener on the jubilation.

Sharma was later posted to a Jaguar Operation Squadron in Delhi as a flight commander, but hardly flew again after that. Reflecting on the untimely death of first cosmonaut Yuriy Gagarin in an airplane accident, he mused, "The Air Force didn't want a Gagarin-like tragedy visiting me."

Back-up pilot Ravish Malhotra retired from the Indian Air Force having attained the rank of air commodore. He is now the co-founder and chief mentor of the aerospace division of Dynamatic Technologies Ltd., in Bangalore, which makes complex aero-structures, aircraft parts and accessories. He heads up a team of over 150 people working on the Sukhoy Su-30 program as well as some of the indigenous projects such as the Pilotless Target Aircraft and the new HAL Jet Trainer.

Malhotra says, "There's still a sense of awe when people become aware of who I am … [but] at the end of the day, one has to have one's feet on the ground." He is very pragmatic about having not launched into space on 2 April 1984, saying, "That's what life is all about." Even so, he recalls being feted and followed by the public gaze. "There was a tremendous reception," on his return to India, he reflected. "The whole thing was televised."

In 1985 Malhotra was awarded the Kirti Chakra, an Indian military decoration, and went on to command a squadron and a base. He was offered the air attaché's job in Moscow, but didn't take it, as "my children were just joining college". He took early retirement in 1994 and began working for a friend who was starting up a defense business.[15]

There's one question Ravesh Malhotra is often asked. "Since I was on the back-up team, people ask me what it felt like when Sharma was assigned to make the flight. Now that's a difficult question to answer. Only one of us could go and at that time you just accepted it."[16]

Ravish Malhotra and Rakesh Sharma in 1995. (Photo: Times of India)

SHARMA LOOKS BACK

Rakesh Sharma, who now lives in Coonoor in the Nilgiris district of India with his wife Madhu, has retired from active employment and currently serves as the Chairman of the Board for Automated Workflow Pty. Ltd. Despite his brief brush with fame, for Sharma, being the first Indian in space seems almost incidental. "If you ask me whether I miss a space flight now, the answer is 'No'. But if you ask me, 'Will you do it again if there's a chance?' then my answer is a resounding 'Yes'."

Sharma rarely reminisces about the time when he made history. "I'm too much of a realist and I take life as it comes," he stresses. Did he ever dream that one day he would travel into space? "I didn't even know that I'd be a fighter pilot, let alone join the club of 300 men [and women] who have been up there."

Rakesh Sharma sits in the cockpit of Boeing F/A-18 he flew at the 2009 Bangalore Air Show. (Photo: The Hindi newspaper)

Sharma still finds it hard to deal with people who, on recognizing him, give what he calls the "customary stare". He says it's sometimes tough on the family. That includes son Kapil, 21, now majoring in communications in Pune, who was just nine when his father went into space; and daughter Kritika, who is often quizzed about her father's historic feat.

When he looks at the sky at night, what does Sharma think about? "Nothing; absolutely nothing. I am a pilot foremost, and flying has always excited me." One of his unforgettable memories of space remains that of "seeing the Earth bathed in the color of blue". But, as he wryly adds, "The euphoria over the event lasted barely a year".

And how does Sharma feel about his present life? "One has to live in the future, not in the past. Those eight days were great moments but one moves on in life. What matters is keeping your feet on the ground."[17]

Asked whether he is still involved in any way in the Indian space program, Sharma said, "Only in an advisory capacity." In fact, upon reflection, he does not believe that his space trip added all that much to the Indian space program. "It was a hugely symbolic event," he admits. "In real terms, it wasn't as big as it is perceived to be. I truly believe I've not done anything extraordinary. Any air force officer trained the way I was would have done exactly what I did."

But there is a rider to this sense of dispassion. "It would have been nicer if I could have done it using my own country's technology. Any Indian could have gone up in somebody else's launch. I would have perhaps reacted differently had I gone up on an ISRO launch." In a way, this explains why Sharma didn't even write a book about his space flight, despite being approached by publishers. "For me, as a person, the space flight was only a part of my life. Even if I'd written a book, it would have only been a travelogue." He struggles to put these thoughts further into context, but believes that ISRO was not all that keen on putting together a manned mission. Its mandate back then was to ensure that the benefits of space technology benefited the common man. "Mrs. Gandhi must have been under pressure from the Soviets, so when ISRO didn't show much interest the air force was approached. ISRO's role in the mission was minimal. They gave us some preparatory lessons in the experiments that we would be conducting. It's interesting that just a quarter of a century later, ISRO is planning its own manned mission with a non-communist Russia."

Sharma's pre-launch training in the Soviet Union was typical of Cold War era events. There was no forced bonhomie. "It was very ceremonial. Everybody would relate to me only on a need-to-know basis. In front of the media, there was a lot of back-slapping, but otherwise it was cold and business-like. For the Soviets, this [joint flight] had to be done because orders had come from the top. It was a flight that was scientific in content, but it also had a lot of pressing political aims."

It was a different time that catapulted Sharma to fame. Would it have been different had he been launched into space with the Americans instead of the Soviets? "I don't think so … but the Americans are far more focused on aims, and there would've been no such thing as propaganda. Whereas for the Soviets, this trip was all about propaganda."[18]

Today the Nehru Planetarium in New Delhi displays the Soyuz T-10 capsule in which Sharma returned to Earth, along with his space suit and mission journal.

REFERENCES

1. Murali N. Krishnaswamy, "Reminisces of a Space Odyssey," *The Hindu* newspaper, 9 June 2014
2. Pankaja Srinivasan, "The Down to Earth Rakesh Sharma," *The Hindu* newspaper, 4 April 2010
3. Kanak Hirani Nautiyal, *The Times of India* newspaper article, "The World is Not Enough," 10 February 2005
4. Air Marshal Narayanan Menon, "Flying the Sukhoi-7 in Operation Cactus-Lily," Bahrat Rakshak, website at: *http://www.bharat-rakshak.com/IAF/History/1971War/1135-Narayanan-Menon.html*
5. Rakesh Sharma interview conducted by Allen Mendonca for *Harmony – Celebrate Age* magazine, July 2005

6. Svetlana Soldatenkova, *Soviet Weekly* newspaper, 21 April 1983
7. *Ibid*
8. *Soviet Weekly* article, "India and Space Research," issue 14 July 1983
9. Kanak Hirani Nautiyal, *The Times of India* newspaper article, "The World is Not Enough," 10 February 2005
10. Rakesh Sharma interview conducted by Allen Mendonca for *Harmony – Celebrate Age* magazine, July 2005
11. Svetlana Soldatenkova, *Soviet Weekly* newspaper, 21 April 1983
12. *Soviet Weekly* newspaper, article "Salyut-7 prepares to receive guests," issue 1 March 1984
13. *Soviet Weekly* newspaper, article "The Joint Space Flight," issue 20 April 1984
14. *Ibid*
15. "Where Are They Now: Ravish Malhotra," Website: *http://www.epaper.timesofindia. com/Repository/ml.asp?Ref=Q0FQLzIwMDkvMDkvMjAjQXIwMTIwMg*
16. Kanak Hirani Nautiyal, *The Times of India* newspaper article, "The World is Not Enough," 10 February 2005
17. "Cosmonaut on terra firma," *Stephen David in Bangalore*, 15 April 1996
18. Sugata Srinivasaraju, O*utlook* magazine article "Feet on the Ground," issue 19 October 2009

13

A Syrian researcher on Mir

In 1985 an invitation was extended to Syria to train two military airmen as candidates for a flight to a Soviet space station. The Syrian chosen to make the flight would become the first guest cosmonaut from the non-socialist developing world. Initially, this joint mission was probably slated for a working occupancy of the Salyut-7 station, but with the launch of the Mir core into orbit on 20 February 1986 those plans would have been modified to a link-up with this larger facility.

Very few details have ever been released on the training program or the actual objectives of the joint mission, but it is known the four finalists sent to Moscow from Syria were Kamal Arabi, Muhammad Ahmed Faris, Munir Habib Habib and Ahmed Rateb – all of whom had previously received higher flight training in the Soviet Union. Muhammad Faris and Munir Habib were selected as the candidates to train for the mission, and they arrived at the Yuriy Gagarin Cosmonaut Training Center in October 1985.

Muhammad Ahmed Faris (left) and Munir Habib Habib. (Photos: Spacefacts.de)

© Springer International Publishing Switzerland 2016
C. Burgess, B. Vis, *Interkosmos*, Springer Praxis Books, DOI 10.1007/978-3-319-24163-0_13

TWO MEN OF SYRIA

Biographical details on the two candidates are disappointingly sparse, but it is known that Muhammad Faris was born on 26 May 1951 in Aleppo, the largest city in the Syrian Arab Republic, some 190 miles from the capital of Damascus. He attended a military pilot school at the Syrian air force academy in Nayrab, near Aleppo, from 1969. On graduating in 1973 he joined the Arab Socialist Renaissance Party and served as a pilot, aviation instructor and specialist in navigation in Nayrab, in the same unit and squadron as the man who would be his cosmonaut-training partner. He was a veteran of two wars with Israel when chosen as a cosmonaut candidate. At that time, he and his wife Gind Akil, a house-wife, had a daughter Gadil, born in 1979, and a son Kutaiba, born 1981.

Munir Habib was born in the Syrian coastal city of Jablah on 3 September 1953. After completing school he entered a university to study English, but subsequently transferred to the Military Air School in Aleppo in 1969, from which he graduated in 1973. After this he received two years of traineeship in Nayrab, during which time he joined the Arab Socialist Renaissance Party, then served as a pilot and instructor in the Syrian Air Force, bearing the responsibility in his unit and squadron for conducting classes in aerodynamics. By the time he was selected as a cosmonaut candidate in October 1985, he had in excess of 3,400 flying hours.

On 23 August 1986 a news conference was held in Damascus, during which Munir Habib and Muhammad Faris asserted that the joint space flight between Syria and the Soviet Union "crowns the growing relations between the two countries in various fields". They added that this space cooperation "indicates that Syria and the USSR have a joint stand in one trench to confront the dangers posed by the aggressive acts of imperialism and world Zionism". They also pointed out that the date set for the joint flight would be some-time in the second half of 1987. "The first stage of our study has been theoretical, during which we learned theoretical sciences and physical exercises," they reported. "The next stage will be an implementation study during which theoretical sciences will be embodied in preparation for space flight."[1]

CREW ASSIGNMENTS AND TRAINING

During training, it was announced that the prime crew of the Soyuz TM-3 mission would be Aleksandr Viktorenko, Aleksandr Aleksandrov and Muhammad Faris. Although this would be Viktorenko's first space mission, he was to be in command. Aleksandr Aleksandrov (not to be confused with the Bulgarian cosmonaut) had spent five months aboard Salyut-7 as the flight engineer for Soyuz T-9 in June 1978.

The commander of the crew that would train with Munir Habib was Lt. Col. Anatoliy Solovyev, along with flight engineer Viktor Savinykh. Solovyev, who had yet to make his first space flight, was born in 1948 and joined the cosmonaut team in 1976. Savinykh was born in 1940. Selected on 1 December 1978 in the same cosmonaut group as Aleksandrov, he had already completed two space missions as the flight engineer on Soyuz T-4, spending 74 days aboard the Salyut-6 station, and on Soyuz T-13, which included a 169-day mission on the Salyut-7.

The prime Soyuz TM-3 crew of Aleksandr Aleksandrov (top of photo), Aleksandr Viktorenko and Muhammed Faris. (Photo: Spacefacts.de)

The back-up crew of Anatoliy Solovyev (bottom of photo), Viktor Savinykh and Munir Habib. (Photo: Spacefacts.de)

Lt. Gen. Georgiy Beregovoy was chief of the training center, and he was impressed with the progress shown by the two candidates. "We have had to meet with cosmonauts from different countries," he stated. "I think that the Syrians were quicker in mastering Russian. They speak Russian fluently."[2]

For the two Syrians, their daily routine in Star City was scheduled down to practically the last minute. Woken at 7 a.m., there would be a period of exercises followed by breakfast at 9:00 a.m., not at home with their families but in the canteen with other cosmonauts. From 11:00 till 1:00 p.m. they attended classes, then dinner, an hour of rest, more classes, supper, and at 7:00 p.m. they were able to go home for a while. In the evening there were further lessons, and at 11:00 it was time for bed.

When asked if he found the routine boring, Munir Habib said that there was no time to get bored. "No time. We are particularly pressed with time for bringing up our children. But in this we are helped by our wives. They are wholly preoccupied with keeping the house. Our elder children, my son Madyan and Muhammad's daughter Ghadil study at an Arab school in Moscow. The younger sons Rayed and Kutayb go to kindergarten. All of our children have become assimilated here much faster than we – they speak Russian better, and have a lot of friends. In the evening they like to watch TV. Madyan is fond of films about war; he often tells me episodes from them, about how [the Russian] people defended their land against the fascists. These films produced a great impression on him."

When asked the same question by a reporter, Muhammad Faris responded, "We spend the weekends with our families. We go to Moscow, to see the Bolshoi Theatre, the Pushkin Art Museum, the circus. We visit other cosmonauts."[3]

The Mir training simulator in Star City. (Photo: Bert Vis)

OFFICIAL ANNOUNCEMENT

The makeup of the final crew for the mission, the fourth manned spacecraft to visit the Mir space station, was officially announced on 17 December 1986 at a USSR Foreign Ministry press conference. It was prefaced by correspondent Aleksandr Galkin stating that the two Syrian candidate cosmonauts had completed a course of theoretical studies at the Gagarin Cosmonaut Training Center and at the beginning of the new year they would start the real preparations for the flight. Maj. Gen. Aleksey Leonov, at that time the deputy head of the training center, announced that the prime crew would include Lt. Col. Muhammad Ahmed Faris. He would fly with the Soviet mission commander Aleksandr Viktorenko and flight engineer Aleksandr Aleksandrov. Leonov then presented an outline of the joint mission.

"The research cosmonauts of the Syrian Arab Republic – Lt. Col. Muhammad Faris and Lt. Col. Munir Habib – arrived at the cosmonaut training center in October 1985. During that same month they started the general space training program. General space training includes the study of fundamental sciences connected with disciplines such as flight dynamics, control systems, the fundamentals of space navigation, and practical work on simulators, as well as medico-biological training, physical training, and the study of the Russian language. A total of 1,400 hours was spent on this. From September to December they studied the systems of the Soyuz T spacecraft and the Mir orbiting station, and particularly the systems with which they will have to work.

"All in all, to date they have arrived at a level where a crew can be formed and the direct preparation for the space mission can be started. The decision regarding the makeup of the crews was taken yesterday.

"The main crew is made up as follows: commander, Lt. Col. Aleksandr Stepanovich Viktorenko; flight engineer, USSR pilot cosmonaut, Hero of the Soviet Union Aleksandr Pavlovich Aleksandrov; and research cosmonaut, Lt. Col. Muhammad Ahmed Faris. The back-up crew is: commander, Lt. Col. Anatoliy Yakovlevich Solovyev; flight engineer, USSR pilot cosmonaut, twice Hero of the Soviet Union, Viktor Petrovich Savinykh; and research cosmonaut, Lt. Col. Munir Habib Habib."[4]

At this time, some of the experiments for the joint research program were identified. Al-Furat (Euphrates) would use the station's KATE-140 topographical camera to study natural resources and agricultural areas, paying particular attention to Syria. The Bosra experiment would use an apparatus designated Missiya which had been jointly developed by the USSR and Syria to undertake ionospheric research. The Palmira and Kasyun experiments were to study the crystallization processes in weightlessness using the Kristallizator furnace, in this case by creating gallium arsenide crystals and a eutectic alloy of aluminum and nickel (an alloy for which the material constituents melt or solidify at the same temperature). And the biomedical experiments Anketa and Kontrast involved a study of human visual response in weightlessness, and purifying samples of interferon and an influenza vaccine.

PLANS FOR THE MISSION

Valeriy Ryumin, the designated flight director for the joint mission, announced a planned launch date of "22 July, at approximately 06:30 in the morning". He also noted that, "The programs of this individual expedition are rich with various experiments; on natural history, geophysical, technological and medical experiments. Specialists of the

Soviet Union and Syria are taking part in their preparation and also in the elaboration of apparatus. We are planning that before the flight of the Syrian crew, an astrophysical module [Kvant] will be docked with the Mir station and work will be carried out. This will include a complex of scientific research using apparatus installed on it. This is mainly astrophysics."[5]

When the Kvant module arrived in April 1987 it was unable to achieve a hard docking. With the module loosely connected to the rear port of Mir in the soft-docked configuration, the crew of Yuriy Romanenko and Aleksandr Laveykin donned their space suits to conduct an emergency EVA. On probing the small gap between the two vehicles with their gloved fingers they found – to their amazement – that the problem was a plastic trash bag that had been left behind when a previous crew was stowing trash into a Progress ship at that same port. With the blockage removed, the module was able to complete its docking. It carried vital equipment, such as Soviet X-ray and gamma-ray telescopes, as well as British, Dutch and German instrumentation.

Amongst other questions at the news conference was one directed to the back-up crew, asking how they felt about not flying. Back-up commander Solovyev responded, "Our crew will be ready in such a manner that it will be fully prepared and will be found to be as good as the first crew." Munir Habib said, "I will behave as a member of the back-up crew just as if I were a member of the main one. Moreover, situations do occur when the back-up crew becomes the main one."

Aleksey Leonov summarized the situation, "All back-up cosmonauts, representatives of countries who flew on our spaceships, haven't broken with space; on the contrary, they have become great specialists, scientists in that area. At present anyone of those who were back-ups in the past, in general, in our view, can take up the place in a spaceship according to his level of training. Two representatives of France – Jean-Loup Chrétien and Michel Tognini – are currently in the Soviet Union undergoing training. They have already started training. At the beginning of January two representatives of Bulgaria will arrive in the Soviet Union. One of the candidates for this place is Aleksandrov, a former research cosmonaut and back-up. Of course we would like for him to pass the repeat medical competition and to join this group."

The two Syrians then told the journalists about life and work in the Gagarin Cosmonaut Training Center and expressed gratitude to its specialists "for their help in coming to grips with a complicated profession. The forthcoming joint work of scientists and specialists and cosmonauts under the two flags of the USSR and the Syrian Arab Republic will become an important landmark in the peaceful mastering of space."

Gen. Abdallah al-Zurf, commander of a Syrian Air Force division, weighed in with his own comments on the selection process.

"We have brought from our country the warmest wishes to the Soviet people and the hope that the Soviet-Syrian crew will successfully fulfill its missions, which we consider to be a bright material expression of the friendship and cooperation which links our countries." He also stressed that "usually the candidate of the main crew is designated by the country from which the cosmonauts come, but we value equally highly both Muhammad Faris and Munir Habib; therefore, we offered the right of choice to the Soviet side. They have more accurate criteria for appraisal."[6]

All is in readiness for the launch of the TM-3 mission to Mir. (Photo: Spacefacts.de)

SYRIA IN SPACE

The Soviet-Syrian joint mission was launched on 22 July 1987. As Soyuz TM-3 roared into space from the Baykonur Cosmodrome for a rendezvous with Mir, Faris was taking with him some dirt from Damascus, a sample of Syrian wheat, and a relic from the dawn of civilization – a tablet which had been discovered in Ugarit, north-west Syria, that was inscribed with the world's first alphabet.

The flight plan was somewhat different to those used in the past; the Soviet planners took advantage of the relatively new 48-hour trajectory between launch and docking. According to Viktor Blagov, a deputy flight director, this reduced the strain on both the flight crew and the ground services, and, more significantly, would be more efficient in terms of propellant consumption. Also for the first time, the spacecraft would approach to dock with Mir at the aft port of the Kvant astrophysical module.

As envisaged by the flight program, on 24 July the Soyuz TM-3 spacecraft docked with the Mir/Kvant/Soyuz TM-2 orbital complex while crossing the Soviet-Mongolian frontier. Radio Moscow said millions of people in the Soviet Union and Syria, which was celebrating sending its first citizen into space, watched live transmissions of the docking. Initially there was no indication of any problem during the link-up. Waiting for them were the residents Yuriy Romanenko and Aleksandr Laveykin, who had been onboard for almost six months. But gaining access into the station was unexpectedly beset with problems, when the Soyuz crew was unable to open the hermetically sealed forward hatch of their vehicle. In the end they had to resort to physically levering it open, inwards against the cabin pressure that was sealing it tight.

Lift-off for the Soyuz TM-3 mission. (Photo: Spacefacts.de)

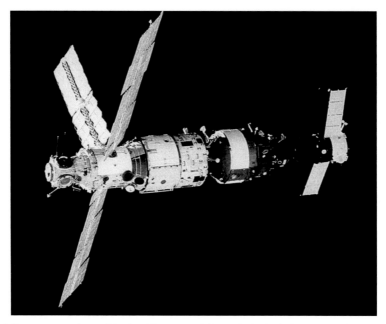

The Mir space station with the Kvant module and docked Soyuz TM-3 spacecraft. (Photo: Spacefacts.de)

Shortly after the two crews had concluded their celebrations on Mir, there was troubling news from the ground. Physicians monitoring telemetry transmitted from his suit during a spacewalk had come to suspect that Aleksandr Laveykin might have developed a potentially serious heart problem, which required bringing him home at the end of the six-day docking mission and replacing him with the newly arrived Aleksandr Aleksandrov. Speaking of this abnormal electrocardiogram, Viktor Blagov told a news conference in Moscow, "It may be serious, it may not be serious."

Laveykin was reluctant to return home, but officials decided to perform the crew switch while they had the chance, lest the condition of the 35-year-old cosmonaut on his first flight deteriorate later and require the station to be vacated. "Although he says he is not tired and believes he can work further, this is the time to take him off," Blagov said, adding that the Soyuz TM-3 mission, which was scheduled to spend six days linked to the Mir conducting medical and scientific experiments, would continue as planned. However, at its conclusion Laveykin would join Aleksandr Viktorenko and Mohammed Faris aboard the Soyuz TM-2 spacecraft for the return to Earth. One can only wonder at Aleksandrov's reaction to being informed that instead of returning in six days, he would remain on board with Romanenko until the next crew arrival ten months later.

When asked about Laveykin's condition, Igor Goncharov of the Institute of Biomedical Problems (IBMP) in Moscow told reporters that telemetric data beamed back to Earth from the spacecraft indicated that all the cosmonauts were in good health.

Reflecting the emphasis that both the Soviets and Syrians have placed on the joint space mission, Soviet leader Mikhail Gorbachev sent the orbiting cosmonauts a message hailing their flight as "a new step in the many-faceted and fruitful cooperation of the Soviet Union and Syria."[7]

Once all of these difficulties and future plans had been sorted, the five cosmonauts began working together. Their tasks included a series of measurements for the Bosra experiment, designed to obtain new information on the physical processes at work in the upper layers of the Earth's atmosphere and ionosphere. The Ruckey apparatus in Kvant, was used to purify genetically engineered interferon and a drug to combat influenza. And the Svetlana facility aboard the new module was to isolate active microorganisms producing antibodies for use in stock farming.

The results of the Euphrates experiment were eagerly awaited in Syria. While flying over that country, the cosmonauts carried out visual observations, took pictures, and performed a spectrometric survey. The data would more accurately define the country's agro-resources, and be of assistance to both geologists and hydrologists.

One of the tasks that required special and unique care involved the Kristallizator furnace. It had been realized that even during designated rest periods, movements by the cosmonauts could create sufficient vibrations to affect the delicate crystallization process. In the case of Mir, experimental work would not be carried out in the space station itself, but in one of the various specially equipped experimental modules docked with it.

Glossed over at the time, there had been health problems on board the Mir station. In a March 2015 interview in *Novosti Kosmonavtiki*, Aleksandr Viktorenko noted Faris was as good as incapable to work because he couldn't, or wouldn't, use the toilet on board Mir for solid functions. Reportedly, he'd only come out of his sleeping bag when he had to appear before the cameras. In the end, Romanenko suggested he take some pills. These seemed to help him, so it would appear that it was a more a case of constipation, which means that he couldn't rather than wouldn't use the toilet.

An essential planned duty for the cosmonauts was to convert the two-man Soyuz TM-2 spacecraft to enable it to carry three occupants to Earth, leaving the fresher vehicle for the eventual return of Romanenko and Aleksandrov. Laveykin, of course, was to return early. This entailed removing Romanenko's customized seat liner from the two-man vehicle and installing it in the descent module of Soyuz TM-3, while the seat liners of Viktorenko and Faris were moved to Soyuz TM-2, which already contained Laveykin's seat liner.

A HITCH ON THE WAY HOME

On 29 January, with their work completed and with experiments and data loaded into Soyuz TM-2, Viktorenko, Laveykin and Faris strapped themselves in and prepared their vehicle for the undocking procedure. There was a frustrating hitch when they were advised that due to high winds and rain squalls in the landing area, they were to delay the undocking procedure for a further two orbits.

Three hours later than planned, Viktorenko undocked from the orbiting complex and soon thereafter the three cosmonauts were hurtling back to Earth. The high winds that had delayed their departure were still quite strong as they descended to the ground by parachute, and their capsule was blown a considerable distance from the planned touchdown point.

Laveykin, Viktorenko and Faris after landing. Back-up cosmonaut-researcher Munir Habib is being interviewed behind Faris. (Photo: Spacefacts.de)

Amid a cloud of yellow dust, the Soyuz TM-2 capsule touched down safely an hour after dawn at 5:04 a.m. Moscow time, 88 miles north-east of the town of Arkalyk in Kazakhstan. Helicopters were quickly on the scene. While Viktorenko and Faris remained in their seats, physicians ran some quick checks on Laveykin. Once the three men were out, they were all reported to be feeling well. After participating in a news conference in the field where they had landed, the crew was flown to Arkalyk by helicopter, and on arriving at the Baykonur Cosmodrome two hours later they were given the traditional welcome of bread and salt.

Laveykin's suspected heart condition, whilst a cause for much concern at the time, later proved to be a very minor ailment. He was eventually cleared for further missions, but he would see out his career as a cosmonaut without making another flight and retired from the corps in 1994.

After his space flight, Faris returned to the Syrian Air Force and settled in his home town of Aleppo with his wife Gind Akil and on 30 December 1987 had a third child, a second son with the unmistakably space-related name of Mir.

For his accomplishments as a cosmonaut, Faris was awarded the title Hero of the Soviet Union and also the Order of Lenin, which was the Soviet Union's highest civilian decoration, and the Order of Friendship of Peoples. He also became a Hero of the Syrian Arab Republic, was awarded the Order of Military Glory, and was appointed an honorary citizen of the town of Arkalyk in Kazakhstan.

Munir Habib, who assisted in the control center in the Moscow suburb of Kaliningrad during the Soviet-Syrian mission, returned to the Syrian Air Force and was awarded the Soviet Order of Friendship of Peoples and the Syrian Order of Military Glory. Starting

Muhammad Faris prior to his defection from Syria. (Photo: Wikipedia)

in 2001 he headed an institute that sought to improve the professional skills of the people of Aleppo. He and his wife Yumna have two sons: Madyan, born in 1976, and Rayed, born in 1981.

On 5 August 2012, reports from Turkey's state-run Anadolu news agency said that 61-year-old Gen. Muhammad Faris had defected from war-torn Syria and crossed into Turkey overnight, where he had joined the opposition forces fighting President Bashar al-Assad's regime. Prior to crossing into Turkey, he had reached the headquarters of the Free Syrian Army in his home town of Aleppo, met with rebel commanders, and declared his solidarity with the group of rebel fighters. "We are with you with our lives and our blood," Anadolu quoted Faris as telling members of the Free Syrian Army. The news agency also reported that it was Faris's fourth attempt to defect after three earlier failures. He joined a string of high-profile figures, including senior military officers, who had abandoned Assad's regime since the start of the uprising in March 2011 and as the fighting escalated in both scope and brutality.

Meanwhile, Faris's whereabouts in Turkey were being kept a closely-guarded secret. It was a propaganda blow for the al-Assad regime, as Gen. Faris was the only Syrian, and one of very few foreigners, to receive the highest Soviet award, Hero of the Soviet Union. It was an honor that had not been bestowed on anyone in the Arab world since Nikita Khrushchev awarded it to President Gamal Abdel Nasser of Egypt and former Algerian president Ahmed Ben Bella. Having joined the insurgents, Muhammad Faris is now serving as a commander in the Free Syria Army.[8]

Munir Habib. (Photo: Spacefacts.de)

Today, the Soyuz TM-3 spacecraft in which Faris was launched into space is on display outside the October War Panorama (or Trisheen) Museum in Damascus, which was built to commemorate the 1973 October War between Israel, Egypt and Syria. Enclosed in a glass case, it sits alongside captured Israeli tanks, some shot-down airplanes, and several Soviet aircraft used by the Syrian military in that conflict.

REFERENCES

1. Text from Damascus SANA state-run news agency (in Arabic), broadcast 1345 GMT on 24 August 1986
2. Svetlana Soldatenkova, article, "The 12th international crew," *Soviet Weekly*, 22 January 1987

3. *Ibid*
4. USSR Report (Space), Foreign Broadcast Information Service, JPRS-USP-87-001, "Cosmonaut crews for USSR-Syria flight presented," 19 February 1987
5. *Ibid*
6. *Ibid*
7. John-Thor Dahlburg, The Associated Press article, "Heart problem forcing cosmonaut home," 24 July 1987
8. Suzan Fraser (Associated Press), "Rebel astronaut? Syria's first man in space defects, says Turkish report, *Christian Science Monitor*, 5 August 2012

14

Bulgaria's second flight

On an official visit to the Soviet Union in 1986, Bulgaria's Minister of the People's Defense Army, Gen. Dobri Dzhurov, reached an historic agreement with the Soviet government that would send a second Bulgarian cosmonaut into space to conduct science experiments aboard the recently launched Mir space station. The Director of the Central Laboratory for Space Research Institute, Professor Boris Bonev, was then sent to Moscow to negotiate follow-on conditions with the newly established Soviet space organization, Glavkosmos.

A NEW COOPERATIVE SCIENCE MISSION

Although similar research plans involving Georgi Ivanov had been thwarted following the docking failure of the Soyuz-33 mission in 1979, the new Soviet-Bulgarian partnership flight would not fully operate under the flag of Interkosmos; rather it would serve as a cooperative science mission. The agreement became official on 22 August 1986 with the signing in Moscow of papers by Aleksandr Dunayev, representing Glavkosmos, and the vice-president of the Bulgarian Academy of Sciences, Mako Dakov, who also served as chairman of the National Committee on the Exploration and Use of Outer Space. These papers detailed the training, preparation and execution of the joint Soviet-Bulgarian space mission. Bulgaria's financial arrangements required that they cover the cost of the joint flight by agreeing to the construction of a large amount of science equipment worth around 14 million U.S. dollars, whose ownership would be transferred to the USSR. The technology needed to make this equipment was to be the responsibility of the Soviet Union.

The candidate selection process began in November 1986. Following a comprehensive medical evaluation of over 300 Bulgarian Air Force officers, 10 were chosen for even more stringent examination by Soviet physicians who arrived in Sofia to conduct the tests. These examinations brought the number of candidates down to just four. One unlucky candidate was Georgi Ivanov, who had previously flown on the luckless Soyuz-33 mission. The four finalists were:

- Aleksandr Panayotov Aleksandrov
- Plamen Panayotov Aleksandrov (younger brother of Aleksandr)

© Springer International Publishing Switzerland 2016
C. Burgess, B. Vis, *Interkosmos*, Springer Praxis Books, DOI 10.1007/978-3-319-24163-0_14

- Nikolay Raykov
- Krasimir Mikhailov Stoyanov.

Following intense medical examinations in Moscow, the Soviet doctors rejected Plamen Aleksandrov for health reasons, and the Bulgarian contingent would make their selection from the three remaining candidates. According to the Bulgarian press, the choice was only made after some intense backroom struggles. This is evidenced by the fact that the official announcement of the two finalists – Aleksandr Aleksandrov and Krasimir Stoyanov – was only made on 5 January 1987, despite the fact that the Soviets had established a deadline of 10 January for the arrival in Moscow of the two finalists. They would be trained to serve as cosmonaut-researchers, which was the customary title assigned to Interkosmos cosmonauts. Losing out in the final selection was Nikolay Raykov, who, like Aleksandrov and Stoyanov, was a pilot graduate from the Georgi Benkovski Higher People's Military Air School.

The two Bulgarian candidates: Aleksandr Aleksandrov (left) and Krasimir Stoyanov. (Photos: Spacefacts.de)

ALEKSANDROV BACK IN TRAINING

Maj. Aleksandr Panayotov (Sacho) Aleksandrov was born on 1 December 1951 in the town of Omurtag, located at the eastern foot of Stara Planina in north-eastern Bulgaria. His father Panayot was a retired worker in the local forestry enterprise, while his seamstress mother had worked in a cooperative until she too became a pensioner. Aleksandr's younger brother also served in the Bulgarian Air Force as an interceptor pilot.

At high school, Aleksandrov's favorite subjects were astronomy, mechanics, electrical engineering, geography, physics, maths and history – all of which would serve him well in his later ambitions. He was also active in many sports, including basketball, gymnastics and swimming.

On finishing his secondary education in 1969, and at the urging of his parents, he applied to the Varna Institute of Mechanical and Electrical Engineering as well as the Georgi Benkovski Higher Air Force Academy. He was accepted by both, but his love of aviation led to him opting for a flying career and he subsequently became a cadet pilot in his nation's air force. Early in his training he was seriously injured when the aircraft he was in crashed into a tree during a parachute jump exercise. After convalescing for several months he was able to return to his studies and rapidly caught up with his colleagues.

Aleksandrov joined the Communist Party of Bulgaria in 1972, and then in the spring of 1974 he received his first assignment as a junior pilot on fighter-bomber aircraft, serving in anti-aircraft units of the Bulgarian Air Force. Within a short time he had mastered all the aspects of his job and greatly increased his piloting skill.

He was selected as one of the two finalists for the first Soviet-Bulgarian space flight in April of 1979 and was teamed with veteran cosmonaut Yuriy Romanenko. However they ended up as the back-up team to Nikolay Rukavishnikov and Georgi Ivanov, who flew the Soyuz-33 mission which failed to dock with the Salyut-6 space station. Now promoted to captain-engineer, Aleksandrov hadn't given up on his ambition to fly in space. In 1983 he graduated from the Institute of Space Research of the Academy of Sciences of the USSR, receiving a science degree in Technical Services. Then upon his return to Bulgaria he was appointed deputy director of the Central Laboratory for Space Research at the Bulgarian Academy of Sciences. The following year he began extramural studies at the G.S. Rakovski Military Academy in Sofia. In late 1986 he was awarded the distinction and title of senior research associate 2nd class. It was around this time that the process began to find a pair of candidates for the second Bulgarian space mission.

Aleksandrov's wife Blagovesta was a graduate of a language high school; at the time of her husband's selection she was studying at the Karl Marx Higher Institute of Economics in Sofia and later became a language lecturer. They have two sons, Panayot, born in 1985, and Radoslav, born in 1989.

Krasimir Mikhailov Stoyanov, whose first name translates to "beautiful world," was born on 24 January 1961 in the ancient north-eastern port town of Varna, on the coast of the Black Sea. His father was a signalman at the harbor's thermo-power station, while his mother was an electric truck driver. Krasimir's brother Veselin became a well-known writer in Bulgaria.

After his eight-form education Stoyanov studied shipbuilding at a secondary vocational school, most likely influenced by the fact that there were a number of naval officers in his family. In subsequent years, however, he became keen on aviation, enrolling in a course in gliding at the Varna aviation club. He completed the theoretical section of his instruction, which confirmed in him a decision to pursue a career as a pilot. In the autumn of 1979 he enrolled in the Georgi Benkovski Higher Air Force Academy.

Stoyanov then married his fiancée Lyudmila, an assistant pharmacist. Between 1984 and 1986 they had two children; a daughter named Mikhaela (1985) and a son Dobromir (1986).

Promoted to the rank of lieutenant-engineer in the spring of 1984, Stoyanov received his first appointment as an interceptor pilot. His initial service with the Bulgarian Air Force began in the 3rd Squadron (IAE) of the 15th Fighter Aviation Regiment (IAP), based in the Black Sea town of Balchek. In the course of his two years he demonstrated excellent results in battle training, piloting skills, tactical expertise and other disciplines which determined his later selection of as cosmonaut candidate. On one occasion he was flying with an instructor pilot in a new aircraft type when they found themselves in what was subsequently described as a "critical situation", and it was noted that Stoyanov handled this undisclosed issue with calmness and expertise.

The prime and back-up crews in a posed photo (Photo: Author's collection)

Commander of the back-up crew Vladimir Lyakhov with Bulgarian crewmember Krasimir Stoyanov in February 1988. (Photo: Author's collection)

According to British space flight researcher Gordon Hooper, "Aleksandrov and Stoyanov were sent to begin their training at the Yuriy Gagarin Cosmonaut Training Center on 10 January 1987 .... At the time, Aleksandrov was a major.

"In November 1987, Aleksandrov was pictured practicing splashdown procedures with Vladimir Lyakhov and Aleksandr Serebrov, and it was assumed that this was the crew that would fly to Mir. However, when the prime crew for the mission was named in December 1987, it consisted of Aleksandrov, Anatoliy Solovyev and Viktor Savinykh. Lyakhov and Andrey Zaytsev were assigned to the back-up crew with Stoyanov, although Serebrov later replaced Zaytsev."[1]

According to Serebrov, it was decided to remove Zaytsev from the crew "for disciplinary reasons."[2]

Splashdown procedures for the Soyuz TM-5 crew. (Photo: Spacefacts.de)

BULGARIA'S SECOND FLAG IN ORBIT

On Tuesday, 7 June 1988, Bulgaria became the sixth nation to send a man into space more than once. Some significant changes had occurred in the flight schedule when the original launch date of 21 June was brought forward in March to 11 June. In April it was further advanced to 7 June. These changes occurred because after the Progress-36 cargo ship had docked at the Mir station in May its engine had been used to modify the trajectory of the orbital complex, and this had impacted on the launch parameters of the ensuing manned mission. In preparing the mission plan, the research program was revised to take this into account.

One by one, nine Bulgarian devices had been added to the permanent equipment aboard the station, and a week ahead of the revised launch date for Soyuz TM-5 these had all been installed and were ready for use. In fact, onboard tests indicated that the Bulgarian devices would likely do an even better job than anticipated. As Sergey Tasalov, a senior expert with Glavkosmos, told the propaganda people, "With its computerized devices Bulgaria holds a leading position among the participants in Interkosmos, the socialist countries' international space exploration program." His boss Aleksandr Dunayev was even more glowing, saying, "By this flight we achieve [a] new quality of space research."

With the American space program still in a prolonged hiatus following the loss of shuttle *Challenger* and her crew in January 1986, the Soyuz TM-5 mission to Mir would be the first manned space mission of 1988. In a nod to an iconic moment in Bulgarian-Russian relations, the Aleksandrov flight was named *Shipka*, a reference to the Balkan Mountains peak where in 1877–1878 Russian and Bulgarian forces won a crucial victory against the Ottoman army in a major step towards Bulgaria's liberation.

Soyuz TM-5 on the launch pad, and lift-off on 6 July 1988. (Photo: Spacefacts.de)

Soyuz launches were usually only broadcast on Soviet television once it was known they were successful, and then only on tape. This time the launch at 6:03 p.m. Moscow time was carried live. The coverage was unusually extensive, lasting more than an hour and featuring live images of the crew in the moments before and after lift-off. Aleksandrov's parents and other members of his family were shown watching the launch on television in Omurtag, and there were scenes of crowds on Bulgarian streets cheering and waving at the news. "I feel excellent," Aleksandrov said several seconds after launch.

As reported several years later by Clive Levlev-Sawyer of the *Sofia Echo* newspaper:

> In line with time-honored practice on both sides during the Cold War, there was the obligatory chat with the head of government, and so viewers of Bulgarian National Television were treated to the sight and sound of long-time dictator Todor Zhivkov, in a hall of the National Palace of Culture, NDK, made up with logos and slogans lauding cosmonauts and communism, in stilted conversation with his compatriot far above. Zhivkov, whose conversation ran to banalities at best, sits in the expansive hall … Aleksandrov stands with his comrades aboard the Soyuz. There is a forced jollity,

a patriotic reference to the "space food" manufactured in Bulgaria, an assurance … that "everything is calm, Comrade Zhivkov". The forced jollity is especially unnerving. All Zhivkov, clad in an ill-fitting sky-blue suit, lacks is a fluffy white cat.[3]

The latter remark was undoubtedly a reference to Ernst Stavro Blofeld, an evil villain in James Bond movies whose ambition was to take over the world.

MISSION TO MIR

The rendezvous would take two days. Already aboard the Mir complex were Vladimir Titov (no relation to veteran cosmonaut German Titov) and Musa Manarov, who had both occupied the station since 23 December the previous year. During their 10-day mission, it was planned for Aleksandrov and his four Soviet colleagues to carry out over 40 experiments, many using the recently installed Bulgarian equipment. The experiments involved space physics, remote sensing of the Earth, and space biology and medicine. The Soyuz spacecraft was carrying an orchid that had been grown on Earth from seeds originally planted on the Mir station.

The five cosmonauts aboard the Mir space station. Bottom (from left): Anatoliy Solovyev, Vladimir Titov and Musa Manarov. At top, Aleksandr Aleksandrov and Viktor Savinykh. (Photo: Spacefacts.de)

Using the newly developed Kurs approach system aboard Soyuz TM-5 (with Mir itself not involved in any maneuvers) the Soviet spacecraft flew around Mir to the vacant docking port at the aft of the Kvant module. Docking occurred at 5:57 p.m. Moscow time on Thursday the 9th. Amongst those shown applauding this success at the control center in Kaliningrad, a suburb of Moscow, was Bulgarian Georgi Ivanov, whose mission nine years earlier had ended in disappointment. About 90 minutes later, Soviet television broke into its evening news program *Vremya* for a live color transmission showing Aleksandrov, wearing a dark blue flight suit, aboard Mir with the four Soviet cosmonauts.

The Soyuz TM-4 descent module prior to landing. (Photo: Spacefacts.de)

A post-landing photo of the Soyuz TM-4 spacecraft, signed in chalk by the returning TM-5 crew. (Photo: Spacefacts.de)

On this occasion, the visitors were to depart in the spacecraft that was already docked at the station, and leave the new one for the residents.

The second Soviet-Bulgarian joint flight ended in triumph at 2:13 p.m. Moscow time on 17 June with a successful landing of the Soyuz TM-4 capsule some 126 miles southeast of Dzhezkazgan in Soviet Kazakhstan. In fact Viktor Savinykh later commented that it was the softest parachute landing he had ever experienced. All three occupants, Solovyev, Savinykh and Aleksandrov were said to be in excellent health and spirits.

As reported in a subsequent edition of the British magazine *Spaceflight News*:

> At around 3:00 a.m. GMT, the returning crew had sealed themselves into Soyuz TM-4 then, following a series of safety and operational checks, undocked as planned at 6:18 a.m.
>
> They then conducted a fly-around of the Mir/Kvant complex, filming the exterior for later analysis by ground-based experts wanting to assess how it is withstanding its long exposure to the harsh environment of low-Earth orbit.
>
> The descent module of Soyuz TM-4 brought back most of the scientific results of the flight, including copious notes, samples produced by microgravity materials-processing experiments, photographs of the Earth's surface, and photographs of space for astronomical studies taken by Mir's battery of multispectral cameras.[4]

It was later speculated that the seven days Aleksandrov spent aboard Mir with his Soviet companions was insufficient to completely fulfill the research program of 46 experiments set up by Bulgarian experts. However, the nine Bulgarian devices remained on board Mir to be used by its long-duration occupants and their future visitors.

Aleksandrov was later asked if he had any moments of concern during the flight. "Was I worried? Yes, on two occasions. During lift-off, and, of course, when docking. We had to succeed by all means. We needed to live up to the hopes and expectations of our scientists. After all, they have been waiting for our missions for years, ever since the dramatic flight of Georgi Ivanov who, due to a technical hitch, failed to dock with the station and to carry out the research program.

"Moreover, we had a tightly packed scientific schedule ahead of us. We were flying nine Bulgarian-made systems (as against three prepared for the previous mission) and we had 46 experiments on our program instead of ten. I find it particularly significant that the project in which we succeeded was named after Shipka, the scene of Russian-Bulgarian comradeship in arms 110 years ago. By the joint program we fulfilled in space, we extended the tenor of this symbol."[5]

A MULTITUDE OF AWARDS

On 30 June 1988, the members of the two space crews and accompanying Soviet experts arrived at Sofia Airport, and later faced a maelstrom of meetings, interviews and travel. As Aleksandrov joked at the time, "The schedule here has proven much more demanding than the program we had to carry out during the flight."

Soon after their arrival, the members of the two crews and the various Soviet guests were received by Todor Zhivkov, First Secretary of the Bulgarian Communist Party and President of the State Council. "Each nation and each person reaches a peak in their existence," he told the visitors. "The people of Russia and the people of Bulgaria have one peak they consider common to both, a peak that symbolizes their friendship. That peak is Shipka. You defended your own Shipka with honor." He then awarded the high distinction of Hero of the People's Republic of Bulgaria to Anatoliy Solovyev, Viktor Savinykh, and Aleksandr Aleksandrov. Aleksandrov also received the title Pilot-Cosmonaut of the People's Republic of Bulgaria. The back-up crew of Vladimir Lyakhov, Aleksandr Serebrov and Krasimir Stoyanov each received the Order of the People's Republic of Bulgaria, First Class.[6]

Now a brigadier general, and awarded the USSR's prestigious Hero of the Soviet Union, Aleksandrov's other decorations and awards include the Order of Lenin (1988); the Order of the People's Republic of Bulgaria, First Class; Order of Stara Planina, First Class (2003); Order of Georgi Dimitrov; Merited Pilot of the People's Republic of Bulgaria; and Merit in Space Exploration (2011). He is also an honorary citizen of Omurtag. After his space flight he went on to become deputy director of the Institute of Space Research at the Bulgarian Academy of Sciences, where he is a research scientist.

In Moscow's Kremlin Palace to celebrate the 50th anniversary of the flight of Yuriy Gagarin, Aleksandrov socialized with the first woman in space, Valentina Tereshkova. (Photo: Wikipedia.org)

Krasimir Stoyanov left the Bulgarian Army with the rank of colonel, and now works for the Space Research Institute of the Bulgarian Academy of Sciences, from which he received his PhD. This Institute is working on 20 space-related projects on a five-year contract with the Russian Academy of Sciences.

In 2011 Stoyanov said he was still somewhat hopeful that a trained Bulgarian cosmonaut (he did not exclude himself) would participate in another space mission. "The decision for sending the next Bulgarian in space will be made on the highest state level; this can happen within the next few years," Stoyanov believes. He also felt that Bulgaria could take part in the Mars 500 experiment, a multi-part ground-based experiment

Soyuz TM-5 back-up cosmonaut-researcher Krasimir Stoyanov. (Photo: Spacefacts.de)

simulating a manned flight to Mars. The experiment's facility is located at the Russian Academy of Sciences' Institute of Biomedical Problems in Moscow. A Bulgarian space greenhouse described as one of the greatest Bulgarian scientific achievements, would be placed on board the Mars 500 "ship" to grow food. Stoyanov emphasized that with its unique space greenhouse, Bulgaria is the first nation which managed to grow wheat and vegetables in space, or space-simulated conditions. "A Bulgarian cosmonaut can apply to take part in the first manned flight to Mars, which is expected in 2025–2030. Why not?" Stoyanov said.[7]

According to Professor Boris Bonev, Director of the Space Research Institute of the Bulgarian Academy of Sciences, Aleksandrov's mission to Mir marked the start of a new stage in his country's space research, the results of which were expected to have extensive applications in crop and forest control, geological prospecting, seismic forecasting, the development of new computer technology, applied medical and biological research, and a number of other areas.

Krasimir Stoyanov in November 2014. (Photo: Wikimedia Commons/Okura)

Bulgaria, the sixth country to send a representative into space, has been a spacefaring nation for more than four decades. This began in 1971 when Bulgaria provided an apparatus for the investigation of space plasma. The nation has placed two satellites and almost 200 instruments into Earth orbit, as well as apparatus used in more than 500 international space experiments. Undoubtedly, the pinnacle of the Bulgarian space research was the flight of Aleksander Aleksandrov in 1988. At that time the country supplied 11 separate equipment units, which were then utilized by cosmonauts for many years on Mir and the International Space Station (ISS), including the RADOM-7 radiation monitoring experiment. In this way Bulgaria continued to be a strong contender in the field of space technology.

At the time of Aleksandrov's flight, Professor Peter Getsov, now director of the Institute for Space Research and Technology at the Bulgarian Academy of Sciences, was the chief engineer of the space science program. He was recently asked on Radio Bulgaria for his view on the significance of the nation's two manned space flights.

"For the science and industry of a country, its participation in space flights is like a high ranking in an Olympic tournament," he responded. "Little Bulgaria ranked among the most important countries in the field of space research and technology. I want to mention another fact, that in 1971 we launched into space our first cosmic device and this made us the 18th country in the world to have designed space equipment and carried out its own experiments in space. Bulgaria is the third country in the world that produces space food. In those years when Bulgaria's first astronaut flew into space and before that, we were the only producer of space foods along with the USSR and the USA. Besides the flights of our cosmonauts, these three factors are significant space achievements for our country. All this is evidence of the creative spirit, highly educated population and technological [proficiency] of Bulgarians."

Bulgaria had conducted over 300 experiments in space to that time, and it continued to participate successfully in the international division of space activities. In connection with these first flights, scientific fields such as space physics, remote sensing of the Earth from space, and space medicine and space biology have all been developing in Bulgaria.

"These programs were a major financial injection for Bulgarian science, industry and technology," Getsov continued. "We purchased large amounts of modern devices, advanced machinery and equipment [and] many people received highly qualified training. It is noteworthy that even now many of these devices are fundamental for some experiments in space. The fact that we had a Bulgarian cosmonaut in space gives higher self-esteem to the nation. Every nation needs above all self-confidence to be able to prosper."

Given these impressive statistics, it would be logical to assume that Bulgaria would now be a member of the European Space Agency. Unfortunately, this isn't so. Given the present-day financial situation of research institutes and technology companies, membership in the European Space Agency is rather expensive. Nevertheless, the Institute for Space Research and Technology at the Bulgarian Academy of Sciences is actively participating in the space ventures of European Framework Research Programs. "We have now about 30 joint projects with other countries," Professor Getsov said in conclusion. "Cooperation opportunities today are greater because there are no political constraints. We work with Russian scientists and companies as well as with European, American, Japanese, Indian, etc."[8]

RESULTS OF A SUCCESSFUL FLIGHT

Bulgarian spectrometric equipment has also taken the first high-quality pictures of Phobos, the larger of the two moons of Mars. A month after Aleksandrov's flight, the first of two Soviet unmanned space probes within the Phobos international project – in which Bulgaria was amongst the most active participants – set off for Mars. A second spacecraft followed shortly thereafter. Phobos-1 was lost in deep space after just two months, probably due to a radio command error. Phobos-2 continued on to Mars and went into orbit around the planet in January 1989, beaming back vital data and images. But on 28 March, while maneuvering to attain a tandem orbit with Phobos, which it was to study closely, all communication was lost. The cause of the failure was later determined to be a malfunction from which the on-board computer was unable to recover.

Joining Danish scientists, Bulgaria also participated in the design and construction of the Soviet-French astrophysics *Granat* satellite that was launched on 1 December 1989 from the Baykonur Cosmodrome using a Proton launch vehicle and inserted into a highly elliptical orbit of the Earth to study energies in the range from X-rays to gamma rays.

Before it was intentionally deorbited and burned up in the Earth's atmosphere in March 2001, it would take the Mir complex just one minute to fly over Bulgaria. Yet, that small country of just 43,000 square miles has demonstrated its ability to continue to share in the peaceful exploration of space.

REFERENCES

1. Gordon R. Hooper, *The Soviet Cosmonaut Team, Volume 2: Cosmonaut Biographies*, GRH Publications, Suffolk, U.K., 1990
2. Bert Vis interview with Aleksandr Serebrov, Planetary Congress of the Association of Space Explorers, Berlin, Germany, 30 September 1991.

3. Clive Levlev-Sawyer, article, "Space for Two Bulgarians," *Sofia Echo* newspaper, 1 April 2011
4. *Spaceflight News* magazine, unknown author, "Soviet/Bulgarian joint mission ends in triumph," issue August 1988, pg. 4
5. *Sofia News* newspaper, article, "A crew's eye view," issue 6 July 1988, pg. 5
6. *Ibid*
7. *Sofia News Agency* unaccredited article, "Cosmonauts eager, hopeful for Reboot of Bulgaria's Space Program," published 17 April 2011
8. *Radio Bulgaria Life* article, "Two Bulgarians flew among the stars," published 4 October 2012

15

Afghanistan's cosmonaut-researcher

Abdul Ahad Mohmand, the youngest of six children for Mohammad Sarvar (of the Pashtun ethnic group) and Sahib Jamal, was born sometime around the first day of 1959 in the city of Sardah, situated in the medieval eastern Afghan province of Ghanzi. The birth was never registered and as his farming parents were unable to recall the exact date they chose the first day of January. His father passed away when Mohmand was in his eighth class at school. Although his mother was illiterate she was well versed in religious matters. However, he found even greater fascination in the skies above.

"When I was little, I used to look up at the sky," he once recalled. "Sometimes planes would appear overhead and I would think how great it would be if I could fly."[1]

TRAINING FOR THE SKIES

In 1976, at the age of seventeen, he traveled to Kabul to attend the Polytechnic University. After graduating the following year he joined the People's Democratic Party of Afghanistan (PDPA), was conscripted into the Afghan air force in 1978 and sent to the Soviet Union for pilot training, first at the Krasnodar Higher Air Force School and then the Kiev Higher Air Force Engineering School. He returned home in 1981 and served for a time in the nation's air force, first as a pilot, then as a wing commander, and later as a chief navigator. He was sent back to Moscow in 1984 for further education, graduating from the Gagarin Air Force Academy in 1987, by which time he was fluent in Russian. On returning to Afghanistan he was promoted to captain and made a deputy regional commanding officer.

On 20 July 1987 an agreement was reached between President Mikhail Gorbachev and General Secretary Najibullah of Afghanistan to fly an Afghan cosmonaut to the Mir space station. The agreement was formalized two months later on 30 September, by Mohammad Watanjar, Member of the Political Bureau of the PDPA Central Committee, and Aleksandr Dunayev, the head of Glavkosmos. The flight was scheduled for the first half of 1989. A commission was set up to identify potential candidates within the Afghan Air Force, which eventually resulted in a pool of 457 names being submitted for further consideration.

In the interview conducted with him, Mohmand was asked when he knew that the hunt was on for an Afghan cosmonaut, and if he was approached at any stage.

© Springer International Publishing Switzerland 2016
C. Burgess, B. Vis, *Interkosmos*, Springer Praxis Books, DOI 10.1007/978-3-319-24163-0_15

"Well, many people were interested in becoming a cosmonaut," he responded. "For me, it was just a fantasy. Then there was an announcement by the supreme command of the Afghan Air Force that said a commission was checking pilots for the position of cosmonaut. I wasn't approached personally. A commission visited all the military units and all the candidates on their list were medically screened and then asked whether or not they would like to become a cosmonaut. I was on a military base in the vicinity of Kabul. I told them, 'Yes.'"[2]

As explained to Mohmand by the commission, the principal requirement was good health, but they were also interested in a candidate's educational background and whether they had a solid knowledge of the Russian language. With only eighteen months' training ahead of the two successful Afghan candidates, there was no time to teach them another language to the necessary fluency. Mohmand said there were some fifty candidates after that first round of interviews and examinations.

"After the second round there were twenty-four, and after the third round there were eight left. After the fourth stage, in Moscow, there were only two left: Dauran and I. That was in January 1988."[3]

Mohammad Dauran (left) and Abdul Ahad Mohmand. (Photos: Spacefacts.de)

The other candidate, Col. Mohammad Dauran, was born on 20 January 1954 in the Nijrab district of Afghanistan's Kapisa province. He attended a military preparation school and then joined the Afghan armed forces in 1972, aged eighteen. Following training in the USSR he served as a MiG-21 pilot and deputy squadron commander with government forces during the Afghan war. At the time of his selection he was a Regimental Commanding Officer in the Afghan Air Force and, like Mohmand, spoke fluent Russian. Dauran's father, a rural teacher, two brothers, and an uncle were all killed during the civil war in Afghanistan.

Did Mohmand know Dauran before their selection as cosmonaut candidates? "Yes, I knew Dauran, he was a commander. Also, from my unit, there were another four or five pilots that I knew." As confirmed by Mohmand, the other six candidates were Shere Zamin, Akar Jan, Mohammad Jahid, Amer Khan, Khyal Mohammad and Syra-Juden. "After the last medical commission," Mohmand continued, "there was a meeting in this medical institute, in which all doctors participated, together with an Afghan delegation. This meeting was very military. The institute was military, and so was I. We were called in one by one, and while we stood [at attention] in front of the commission we were informed of the outcome of the screening. They told us what they had found, what they had not found [and] whether you were in good health or not. After that, they would say something like, 'And therefore, you can, or cannot, become a cosmonaut.' The chairman asked me, and probably also Dauran, 'Do you want to do this … do you want to become a cosmonaut, or not?' I told him, 'Yes.' And so Dauran and I started our training."[4]

REPORTING FOR DUTY

The names of Mohmand and Dauran were formally released on 12 February 1988. Amid their excitement, as Mohmand has revealed, there was one pre-notified requirement for the chosen pair to fulfill – the day after the commission they both underwent a tonsillectomy. "That had been one of the points of the commission. They told both Dauran and I that we needed a minor operation. We didn't even have time for a party."[5]

The prime crew of Mohmand, Lyakhov and Polyakov. (Photo: Author's collection)

The back-up crew of Berezovoy, Dauran and Arzamazov. (Photo: Spacefacts.de)

The two candidates arrived in Star City on 25 February and began training by themselves the following day, starting with the basic theory of space flight. Next they moved on to survival training in a deep forest, learning how to doff their spacesuits and put on warm clothing until a helicopter rescue could be effected. By this time 29-year-old Mohmand was training with his Soviet crew of mission commander Vladimir Lyakhov and Dr. Valeriy Polyakov (who would remain aboard Mir for approximately a year monitoring the health and well-being of himself and two resident crews), while Dauran underwent similar training with cosmonauts Anatoliy Berezovoy and German Arzamazov.

In 1993, Yuriy Malyshev revealed there had actually been a third, all-Soviet crew assigned to the Soyuz TM-6 mission, comprising himself and Aleksandr Borodin. "This crew was the reserve crew," he revealed, "only to replace cosmonauts from the first or second crew, should they become disqualified."[6]

Late in the training process the two Soviet-Afghan crews were scheduled to perform sea survival training, but while Mohmand's crew completed this exercise, Dauran had fallen ill and could not take part. Ordinarily the Soyuz crews would progress to the far more severe desert and cold region training fall, but owing to the pressure of increasing time constraints Mohmand and Dauran did not participate.

During this period Mohmand and his wife Bibigul, a journalist for the Afghan newspaper *Hewad*, had a daughter named Hila, born in May 1988.

Aleksandr Borodin from the reserve crew in training. (Photo: Spacefacts.de)

Then the Soyuz TM-6 mission for which they were training was dramatically advanced from July 1989 to August 1988 to ensure that it would occur before the final withdrawal of Soviet troops from Afghanistan, while a Soviet-friendly government was still in place. So the two Afghan candidates ended up receiving only a very basic and somewhat hectic six months of training. Around six weeks before the scheduled launch date, it was officially announced that Lyakhov, Polyakov and Mohmand would be the prime flight crew, with Berezovoy, Arzamazov and Dauran as the back-up crew. Specific mission training for the prime crew now greatly intensified.

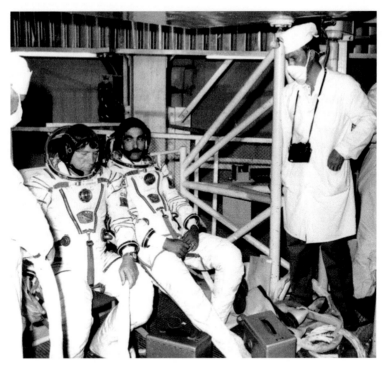

Looking fatigued, Lyakhov and Mohmand rest during a training exercise. (Photo: Author's collection)

Although he was in far better health than Dauran, Mohmand had harbored suspicions that his more senior countryman might have had a political advantage, "because he knew people in the Politburo and the Central Committee. But … the number one factor is health. And physically, I was in better shape. I had no problems running or playing soccer. I swam, I endured centrifuge rides better. I was thinking that I would be chosen if they would look at criteria such as health, but he had many [very influential] acquaintances. There *was* some political pressure for the assignments, but in the end the physical fitness factor prevailed."[7]

Contemporary reports state that the older, more experienced Dauran had been chosen to fly the Soviet-Afghan mission, but Mohmand denies this, insisting that ultimately it was a health issue that saw him assigned to the prime crew – when Dauran's illness proved to be appendicitis he needed an operation. Despite his older, more senior countryman losing the flight, Mohmand emphasized that their relationship was "always good. Always. You know, I was a low-ranking officer, he was a high-ranking officer, and I always told him that he was my commanding officer."[8]

Prior to the scheduled launch date the prime crew was given a three-day break in which to relax, which they duly spent in Moscow.

At left, Berezovoy, Dauran, Mohmand and Lyakhov stand in front of the Soyuz TM-6 on the launch pad. At right, the rocket stands poised for the mission. (Photos: Spacefacts.de)

LINK-UP WITH MIR

Along with mission commander Vladimir Lyakhov and Dr. Valeriy Polyakov, Mohmand was launched aboard Soyuz TM-6 at 04:23 GMT on 29 August 1988. Two days later it linked up with the Mir space station and the visitors made their way inside, where they were greeted by the resident crew of Vladimir Titov and Musa Manarov, then in the ninth month of their long-duration mission.

The Soyuz TM-6 mission took to 208 the number of people who had made space flights to that time – 66 Soviet citizens, 120 American citizens, three West Germans, two French, two Bulgarians, and one each from Czechoslovakia, Poland, East Germany, Hungary, Vietnam, Cuba, Mongolia, Romania, India, the Netherlands, Canada, Mexico, Saudi Arabia, Syria, and now Afghanistan.

During his six days aboard Mir, Mohmand took pictures of his country and participated in astrophysical, medical, and biological experiments on fifteen-hours work shifts. One special task was to brew Afghan tea for his colleagues. At the request of the Afghan government, he was filmed reciting passages from the Koran. "Ahad put a skull cap on to do it," Lyakhov recently recalled. "He was being filmed from below and I was just out of shot hanging on to his legs to stop him floating off."[9]

Mohmand also received a call from Afghanistan's president, Mohammad Najibullah, who arranged for Mohmand's anxious mother to talk to him, so that she could be assured that he was okay.

One of Mohmand's major objectives was a national one, which he said consisted of taking photographs of his native land from the station using different cameras.

"One of the things I did was to photograph the entire territory of Afghanistan. That was of interest to our delegation. They wanted to map the country's resources. Water, for instance, is very important for Afghanistan, and oil, gas. And for the first time we could make an atlas of Afghanistan. Until then, there was no atlas. Whenever the station flew over Afghanistan I would prepare to begin taking photographs. Every time, I was assisted by one or more of the Soviet cosmonauts, so that we could take photographs of the same area at the same time with different cameras."[10]

Musa Manarov proves that there is no real up or down when working in weightlessness. (Photo: Author's collection)

Mohmand also conducted a number of medical experiments, again assisted by the Soviet crewmembers. Repeating such experiments with different visiting cosmonauts enhanced the value of such tests, by increasing the number of cases in the resulting database.

On 6 September, after six days aboard Mir, the time came for Lyakhov and Mohmand to prepare for the journey home. Polyakov, a physician, would remain aboard Mir to monitor the health of the residents and offer psychological support. Lyakhov and Mohmand were to return to Earth in Soyuz TM-5, leaving behind their newer spacecraft.

Problems began shortly after undocking when a computer-controlled engine used to slow the Soyuz spacecraft for re-entry into the atmosphere fired automatically but cut out abruptly as the vehicle crossed the terminator into sunlight. The navigational computer subsequently signaled that it was unable to confirm that the spacecraft was in the correct orientation for a re-entry burn. The problem was traced to a malfunctioning infrared sensor, which measured the heat radiated by the Earth relative to the cold of space in order to orientate the spacecraft ready for re-entry. The Soviet news agency TASS reported that the sensor was incorrectly advising the on-board computers that the spacecraft was improperly oriented. As they had already jettisoned the orbital module of the vehicle, which contained the docking apparatus, they could not make their way back to the safety of the Mir station.

As Soviet cosmonaut Aleksandr Aleksandrov briefed reporters in the control center, "Sun rays prevented the sensor from coming into operation. The computer interpreted that as a loss of orientation and inhibited the switching on of the deceleration motor."[11] A second orienting sensor also malfunctioned, and by the time the cosmonauts were able to override the sensors, they had already traveled too far to touch down in Kazakhstan. They "could have landed on Chinese territory," TASS noted. There were also fears that if they were forced to remain in space much longer they might succumb to a build-up of deadly carbon dioxide. Also, their remaining food rations and the toilet had been in the jettisoned orbital module. So it would be necessary to resolve the dilemma rapidly.

EMERGENCY LANDING

Two orbits on, the next attempt to land was thwarted when the deceleration motor burned for just six seconds instead of the required 230. It was reported that the computer program which controlled the deceleration had not been properly checked and cleared after the first mishap. In fact, it had apparently not occurred to either the crew or the ground team to reprogram the computer for the spacecraft's new position. Lyakhov responded by pressing a manual button to restart the engine, but once again the computer detected an incorrect spacecraft orientation and shut it down after 50 seconds. "I am not excusing myself," Lyakhov would later admit. "There was fault there."

The world's press quickly caught onto the story, and soon thereafter, dramatic headlines hinted that a possible tragedy was in the making. "Trapped in Space: Cosmonauts 48 Hours to Live" declared the front page of the *Daily Mirror* newspaper in Sydney, Australia. It went on, "The Soviet life support system has only enough air left for up to 48 hours and there is no food aboard." James Oberg, a contractor at the Johnson Space Center and well known space historian, said at the time that the on-board failures were probably the result of accelerating the timing of this particular flight. The Afghan-Soviet mission was originally scheduled for the following year but was suddenly moved forward in February, evidently to put an Afghan cosmonaut in space as the aforementioned gesture of Soviet-Afghan friendship before Soviet troops completed their departure from Afghanistan early the next year.

"They were supposed to fly in July of 1989," Oberg stated. "The only way they could fit it in this year was to launch and land early in the morning. Thus the sunrise and sunset times along the orbit were unusual. The initial failure of the infrared orientation sensors is likely to have occurred because Soviet cosmonauts have little or no experience in making deceleration burns just at the day-night boundary."[12] He pointed to the "incredible haste" with which the mission was flown and argued the late-summer launch had led to the disorienting encounter with the Sun's rays. Oberg calculated that the cosmonauts had narrowly avoided falling into an unstable, atmosphere-grazing orbit. Had that occurred, both they and their vehicle would have been completely incinerated.

As Soviet engineers and technicians worked furiously to overcome the problem, Lyakhov and Mohmand were told they would need to spend an extra day in orbit aboard the cramped confines of the descent module.

A ground controller asked Lyakhov, "How are things with food?"

The cosmonaut responded, "There is no food."

"What about the emergency rations?" the controller suggested.

"They are there," Lyakhov replied. "But why touch them? We will be patient," he added, pointing out that there was no way to rid themselves of wastes.[13] Fortunately they were able to temporarily resolve this problem by using the plastic sleeves of Mohmand's sleeping bag as makeshift waste and urine collection bags.

"It was a difficult situation," Mohmand recalled. "TsUP ordered us to remove our space suits. I told them, 'No, we won't do that. It's better if we keep them on.' I wasn't sure that we would be able to land the next day. If we had taken the suits off, we would have had to put them on again the next day and then test them again, to verify that they were airtight. I believed it was better for us to keep them on. Oxygen was a valuable thing and we should save it, just in case we needed to wait *another* day. So I told them we wouldn't do it. All I did was talk with Lyakhov. All the food and drink was in the orbital module … and also the toilet." The orbital module, already detached from the descent module, was long gone and couldn't be retrieved. "And sleep," he added, "we slept a lot."[14]

As time and oxygen supplies ran low, ground controllers in Kaliningrad finally managed to reprogram the Soyuz computers for a new re-entry path. On the morning of Wednesday, 7 September, Lyakhov and Mohmand were given instructions from the ground to make a third attempt at retro-fire.

As explained by Rex Hall and David J. Shayler in their 2003 book *Soyuz: A Universal Spacecraft*, "the revised computer programme was read up to Lyakhov, who then entered it manually into the computer (a procedure that the Americans had used several times during the aborted lunar mission of Apollo-13 in 1970). The orbit had decayed slightly as a result of the various aborted manoeuvres, but by this time using the back-up engine as a precaution, the alignment was perfect."[15]

To everyone's relief, all went well and the retro-fire took place without further incident. Some 26 hours after the emergency began, a large orange-and-white parachute blossomed above the capsule after it had successfully followed a fiery path through the atmosphere. It then floated down and landed a mere six miles from its re-plotted touchdown site in Soviet Kazakhstan. However, there was no live radio coverage of the landing; only live TV from the control center. Cosmonaut Vladimir Solovyov said that the two cosmonauts aboard the spacecraft "displayed good nerve during the incident and had the situation under control."[16]

Lyakhov and Mohmand were later shown sitting outside their capsule being interviewed by reporters. Lyakhov said that it had been uncomfortable without proper toilet facilities but that the situation had always been under control. Mohmand added, quite simply, that these things happen on space flights, and while they had endured a perilous delay to their landing they had been more tense than afraid.

POST-FLIGHT CELEBRATIONS

Mohmand was later awarded the highest Soviet decoration of Hero of the Soviet Union and the Order of Lenin. Lyakhov, who had already received two such awards after earlier flights into space, was only awarded the Order of Lenin. Additionally, Mohmand was awarded the medals of Hero of Afghanistan and the Freedom Sun.

Lyakhov and Mohmand flew to Kabul soon afterwards, receiving a rapturous welcome from cheering crowds lining the city's streets, even as the mujahedin, fighting against the Soviet occupation, were firing rockets into it.

A recent photo of Gen. Mohammad Dauran. (Photo: Spacefacts.de)

Post-flight, Mohammad Dauran was awarded the Soviet Order of People's Friendship. On returning to Afghanistan he was promoted to the rank of general and appointed commander-in-chief of the Afghan Air Force, living in Kabul. In 2009 it was announced that he had been named Afghan National Army Air Corps commander to operate as the counter-part to the U.S. commander of the Combined Air Power Transition Force in Afghanistan. These days he and his wife Alia Nur Mohammad are reported to have six children.

Because it occurred toward the end of the Soviet occupation of Afghanistan, Mohmand's flight to Mir and his status as the first Afghan citizen in space (aboard a Soviet spacecraft) carried significant symbolic importance for a time. Following his one and only mission, he spent the next two years as a student at the Voroshilov General Staff Academy in Moscow, then returned to Afghanistan, where he worked for three months in a space science institute. Meanwhile he found to his annoyance that he had become something of a protected identity and was obliged to work at home. "I was a member of the party and many members of the Politburo and the Central Committee said I was a cosmonaut and that cosmonauts shouldn't work. [Therefore] I was at home and my wife worked."[17]

Abdul Ahad Mohmand in recent times with a model of his Soyuz TM spacecraft. (Photo: DPA)

A year later Mohmand was appointed Deputy Minister for Aviation and Tourism, but this job lasted only six months before the mujahedin swept the Soviet-backed government out of power in a murderous takeover. By this time he had decided that he and his family ought to leave his war-torn country and seek political asylum in Germany, where his brother had lived in peace for thirty years on the outskirts of Stuttgart. "I hadn't seen him in a long time. For me, it was difficult to travel to Germany, and for him it was difficult to travel to Afghanistan. After I arrived [in 1992], I thought the situation might change soon, and I'd be able to return to Afghanistan. But it got worse and worse."[18] When the family arrived in Germany after a hastily arranged 'business trip' they only had a single suitcase of their possessions. He didn't carry anything related to his space flight except his copy of the Koran and a religious rosary. Today, he does not know what became of the spacesuit he wore, or any other mementoes of that time.

Initially, Mohmand had to go on welfare while his family settled in Germany, but he later took part-time work at the University of Stuttgart and went on to work in a print shop, where it took a while for his co-workers to accept that he had been in space. He also mentioned that when he finally received his identity papers from the German authorities, they had made a mistake in spelling his name Momand. He decided simply to leave it at that, and now uses the name Abdul Ahad Momand.

Asked what his prospects were of returning to his homeland one day, he was circumspect and replied, "Very difficult to say. When things quiet down in Afghanistan I will go back. I mean, when Afghanistan has a serious, democratic government and people can live freely and speak out freely, then I'll go back. As long as there is a fundamentalist, radical government, there is no chance of me returning. For my children it would be difficult, from the point of education. I would be forced to wear a turban [and] my wife would be forced to wear a veil and be forbidden to work. She is a journalist."[19]

In December 2013 Mohmand finally returned to Afghanistan, anxious to find out how he would be received. He need not have worried, as he had barely unpacked when he received an invitation to join President Hamid Karzai for lunch at the Presidential Palace. There he finally got the chance to discuss his memories of his flight and present the president with a gift book that had been written afterwards.

In conclusion, Mohmand told President Karzai that "when I reached space I thought that the planet Earth is our common home and it belongs to all the human beings and all should live on it peacefully."[20]

REFERENCES

1. Jenny Norton, *BBC News Magazine* interview with Abdul Ahad Mohmand, published 23 March 2014. Website: *www.bbc.com/news/magazine-26648270*
2. Bert Vis, interview with Abdul Ahad Mohmand, Germany, 20 October 1995
3. *Ibid*
4. *Ibid*
5. *Ibid*
6. Bert Vis, interview with Yuriy Malyshev, Star City, Moscow, 6–15 Aug. 1993
7. Bert Vis, interview with Abdul Ahad Mohmand, Germany, 20 October 1995
8. *Ibid*
9. Jenny Norton, *BBC News Magazine* interview with Abdul Ahad Mohmand, published 23 March 2014. Website: *www.bbc.com/news/magazine-26648270*
10. Bert Vis, interview with Abdul Ahad Mohmand, Germany, 20 October 1995
11. *Sydney Morning Herald* newspaper, article by Felicity Barringer, "Soviet cosmonauts land safely after 25-hour ordeal in ailing spacecraft, Thursday, 8 September 1988

12. *Ibid*
13. *Time* magazine article by John Greenwald, "Close Call over Kazakhstan," issue Vol. 132, No. 12, 19 September 1988
14. Bert Vis, interview with Abdul Ahad Mohmand, Germany, 20 October 1995
15. Rex Hall and David J. Shayler, *Soyuz: A Universal Spacecraft*, Springer-Praxis Publishing, Chichester, U.K. 2003, pp. 329–332
16. *Sydney Morning Herald* newspaper, article by Felicity Barringer, "Soviet cosmonauts land safely after 25-hour ordeal in ailing spacecraft," Thursday, 8 September 1988
17. Bert Vis, interview with Abdul Ahad Mohmand, Germany, 20 October 1995
18. *Ibid*
19. *Ibid*
20. *Bakhtar News*, unaccredited article, "President Karzai Meets Abdul Ahad Mohmand," issue 7 December 2013

16

Chrétien in space again

The year was 1988, and significant winds of change were blowing through the corridors of power in the Soviet Union. In January the program of economic restructuring that came to be known as *perestroika*, with legislation initiated by Soviet leader Mikhail Gorbachev, began to rapidly gain momentum.

In April, after more than eight long and frustrating years of bloody, vicious fighting in an ultimately fruitless occupation of Afghanistan, a commitment was ratified for the withdrawal of Soviet troops from that war-ravaged country. Their departure began the following month. Also in May, U.S. President Ronald Reagan paid an historic visit to the Soviet Union (which he had aggressively characterized as an "Evil Empire" just five years previously) in order to attend a vital and vigorous Moscow summit with Gorbachev.

Meanwhile, in the United States, the launch of space shuttle *Discovery* carried America back into space again following the *Challenger* disaster of January 1986. And in November, Frenchman Jean-Loup Chrétien would complete his second flight into space aboard a Soviet spacecraft, Soyuz TM-7, on a planned 23-day mission aboard the Mir space station. In doing so, he became the only foreign cosmonaut to have worked aboard two Soviet space stations; Salyut-7 and Mir.

A SECOND TRIP INTO SPACE

On his first space flight aboard Soyuz T-6, six years earlier in 1982, Jean-Loup Chrétien had been seated alongside Soviet cosmonauts Vladimir Dzhanibekov and Aleksandr Ivanchenkov on a flight to the Salyut-7 space station. Once there, he performed a number of experiments, several of which were related to human adaptation to weightlessness.

By the strict Interkosmos 'rules' Chrétien's Soyuz T-6/Salyut-7 mission lasted just under eight days, but his next flight would carry him to the orbiting Mir station in November 1988 as part of a 30-day mission, during which he was due to spend about 160 hours working and experimenting with some 1,100 pounds of scientific equipment aboard that station. It would be the longest international flight ever undertaken by the Soviet Union, as well as one of the most ambitious, and it would follow two other international flights to this sophisticated new space station in the same year, involving a Bulgarian and an Afghan cosmonaut-researcher.

© Springer International Publishing Switzerland 2016
C. Burgess, B. Vis, *Interkosmos*, Springer Praxis Books, DOI 10.1007/978-3-319-24163-0_16

French spationaute Jean-Loup Chrétien. (Photo: CNES)

This second Franco-Soviet mission had its origins at a meeting held between the Soviet Glavkosmos organization and the Centre National d'Etudes Spatiales (CNES), the French space agency, in March 1986, and was formally agreed upon by both nations in July. This new flight by a *spationaute* would cost France around $US21 million. The agreement was signed by Dr. Jacques-Louis Lions as the president of CNES, and Academician Vladimir Kotelnikov, vice-president of the USSR Academy of Sciences and chairman of the ruling Council of Interkosmos. In October the flight was officially ratified by Mikhail Gorbachev and French President François Mitterand during the latter's state visit to Moscow.

Earlier, on 1 August 1986, CNES had announced that it had finished screening candidates for the upcoming joint flight, which had been scheduled for late 1988, and that France's first cosmonaut Jean-Loup Chrétien had been selected for the mission.

Michel Tognini. (Photo: Spacefacts.de)

At the time of his selection Chrétien was working at France's National Space Center, taking part in research connected with the development of manned spacecraft. It was further announced that test pilot Michel Tognini, a lieutenant colonel in the French Air Force, had been chosen as Chrétien's back-up.[1]

The CNES candidates who had been evaluated for the flight were Chrétien, Tognini, Jean-Pierre Haigneré and Antoine Couette. Jean-François Clervoy had originally been named as a candidate, but his name was later withdrawn due to a medical problem. Eventually Couette also lost his place as a candidate, and the final selection fell to a choice between Chrétien and Tognini. Both Haigneré and Clervoy would later become astronauts with CNES and later the European Space Agency (ESA), and would fly in space.

Final details of the joint space flight were organized in October 1986 in the Armenian capital of Yerevan, where the official designation 'Aragatz' (named for Armenia's highest summit) was bestowed upon the flight project. Meanwhile, Soviet cosmonaut Aleksandr Volkov had been named as the commander of the Soyuz TM-7 mission, although a Soviet flight engineer was still to be selected. Volkov had flown aboard Soyuz T-14 in 1985 and spent 64 days aboard Salyut-7.

CNES finalists Jean-François Clervoy (left) and Jean-Pierre Haigneré. (Photos: Spacefacts.de)

Chrétien and Tognini were told from the outset that their tenure aboard Mir might last as long as three months, which was exciting news for both men, even though only one of them would make the flight.

In announcing the Franco-Soviet mission to Mir, Aleksandr Dunayev, chief of the Soviet space agency Glavkosmos, said that the most spectacular aspect would be a five-hour EVA (Extra-Vehicular Activity, commonly referred to as a spacewalk) by Volkov and Chrétien, with the latter becoming the first non-Soviet or non-American citizen to participate in this activity. During the spacewalk they would deploy a new structure on Mir's exterior. As detailed by the late space writer Neville Kidger, the new experimental deployment structure was called ERA, and it was "developed by Aerospatiale and comprises a hexagonal-shaped series of carbon-fiber tubes linked together by light alloy hinge joints. An interconnected series of 1 m-long bars form a series of triangles. It is bundled together in a 1 m-long x 0.6 m-diameter cylinder for delivery to Mir.

"During the EVA the bundle will be attached outside one of Mir's front docking ports. When the two cosmonauts return inside it will automatically unfurl to its full 3.8 m-diameter shape in just four seconds, watched by a camera. Tests will then be made of the rigidity of the structure to vibrations. These will be measured by micro-accelerometers. After these tests are completed the structure will be cast off into its own orbit."[2]

The two spacewalkers were also scheduled to perform an experiment program known as Énchantillons, which translates as "samples". It involved mounting a square plate covered with small specimens of paints, coatings and film outside the ship for a series of six-month tests to check the effects of solar radiation, extreme temperatures and other hazards of long exposure to the space environment. It would later be recovered by the crew of Soyuz TM-8 and returned to Earth for analysis. However, apart from saying that the French cosmonaut would conduct medical experiments Dunayev failed to provide more specific details of the program.

TRAINING FOR THE ARAGATZ MISSION

Chrétien and Tognini arrived in Star City on 15 November 1986 to begin their cosmonaut training. Chrétien was accompanied by his family, but Tognini was divorced from his first wife and came alone.

Michel Ange Charles Tognini was born on 30 September 1949 in Vincennes, Val-de-Marne, in the eastern suburbs of Paris. His parents Jean and Ginette (née Choukroun) were both employed by Air France; his father worked as a commercial agent for the airline while his mother was a secretary.

In his youth, Tognini was educated at the College de l'Hay-les-Roses and then at the Lycée de Cachan, Paris. He received an advanced mathematics degree in 1970 from Epa Grenoble (military school). Following this, he enrolled at École de l'Air (the French Air Force Academy) in Salon de Provence, France, graduating with an engineering degree in 1973. Tognini was then posted to advanced fighter pilot training at a squadron based at Normandie-Neman, where he served for a year before gaining his advanced fighter pilot training. From 1974 to 1981 he was an operational fighter pilot in the French Air Force (Cambrai Air Base) at the 12th Escadre de Chasse, flying SMB2 and Mirage F1 aircraft, serving as a flight leader (1976) and a flight commander (1979). In 1982 he attended the Empire Test Pilots School in Boscombe Down, England, and got his diploma in military studies the following year.

Next, Tognini was posted to the Cazaux Flight Test Center in France, initially as a test pilot and later as chief test pilot. During this time, he was involved in testing an impressive amount of French flight hardware. He did the weapon systems tests for the Mirage 2000-C, Mirage 2000-N and Jaguar aircraft, particularly the ATLIS (Automatic Tracking and Laser Integration System) and FLIR (Forward Looking Infrared) pods, and he was responsible for flight safety for pilots, experimenters and flight engineers.

In 1985 France opened a recruitment program to expand its astronaut corps and Tognini was one of seven finalists selected by CNES in September 1985. He remained an air force officer, but under attachment to the space center.

Prior to divorcing, Tognini and his first wife Anne-Marie (née Segura) had two children; Nicolas and Benedicte. While undertaking his lengthy training in Star City he entered into a relationship with physical education instructor Yelena Chechina and on 27 April 1988 they would marry.

In January 1988 Tognini was paired with cosmonaut Aleksandr Viktorenko in the back-up crew for Soyuz TM-7. Viktorenko had commanded the Soyuz TM-3 mission in 1987 which took the Syrian cosmonaut Muhammed Faris to Mir. Neither of the new crews had yet been assigned a Russian flight engineer, but this would change when Aleksandr Kaleri was named to the prime crew and Aleksandr Serebrov to the back-up crew. Two months later, however, Kaleri was temporarily removed from flight status for medical reasons, and he was replaced by Sergey Krikalyov who would be making his first space flight.

The Soyuz TM-7 back-up crew. From left: Michel Tognini, Aleksandr Viktorenko and Aleksandr Serebrov. (Photo: Author's collection)

A SHORTENED STAY

Towards the end of Chrétien's two-year training there was a change to the timeframe for the Soyuz TM-7 mission which caused his already shortened excursion aboard Mir to be further reduced by five days. Late in the preparations, President François Mitterand was invited to attend the launch at the Baykonur Cosmodrome as part of his official visit to Moscow. The original schedule called for the launch to take place on 21 November, but when Mitterand accepted the invitation he was not due to arrive in the Soviet capital until four days after that date, so the launch was rescheduled to 26 November. Chrétien's planned journey home from Mir, however, remained unchanged; he would return to Earth on 21 December with Vladimir Titov and Musa Manarov in the Soyuz TM-6 spacecraft, leaving Aleksandr Volkov, Sergey Krikalyov and Valeriy Polyakov aboard Mir.

Since his arrival in Moscow for mission training, and much to his chagrin, Chrétien had seen the time that he would spend aboard Mir progressively reduced, which he did not find all that satisfactory. "Initially I was supposed to stay three months … Then it was shortened to forty-five days. Then at the end, a couple of weeks before we left, they had to reduce it to one month. Then they had to reduce it again four or five days because the French president couldn't be at the initial launch date. So my flight was shrinking, and for me it was a short flight of one month … It was like vacation when you get half of your vacation, that I wish it was still 100 percent of it, that I was wondering, how can I stay more? But I didn't."[3]

The prime Soyuz TM-7 crew: Jean-Loup Chrétien, Aleksandr Volkov and Sergey Krikalyov.
(Photo: Spacefacts.de)

Chrétien in weightless training for his spacewalk aboard a specially modified jet airplane.
(Photo: Author's collection)

Chrétien and Tognini trained in EVA procedures in a large water tank in Star City. Their spacesuits were made neutrally buoyant to help simulate weightlessness. (Photo: CNES)

Just 11 days before the revised launch date, on 15 November, the upcoming mission was temporarily overshadowed by a major event in Soviet space flight history. Having arrived at Baykonur several weeks beforehand to undertake final mission training, Volkov, Krikalyov and Chrétien and their back-up crew were gently woken by staff members of the Cosmonaut Hotel who advised them that a much-anticipated launch was about to take place.

Soviet shuttle 'Buran' on the launch pad prior to its one and only mission. (Photo: buran.ru)

The weather outside was inhospitable, with gusting winds and temperatures near freezing. After dressing in warm clothes, the six men and a number of hotel staff made their way to the roof and their attention was drawn to a brightly illuminated launch pad in the distance, where the Soviet Union's delta-winged Buran (translating as "snowstorm"), developed in response to America's space shuttle program, stood poised to make its first (and as it turned out, only) autonomous flight into space, coupled to a massive Energiya booster rocket. The white-tiled spacecraft, gleaming under the bright lights, was unmanned and completely controlled from the ground for its maiden, two-orbit flight. It was a cold and windy morning with flurries of snow, and there were indications of a cyclone moving in from the Aral Sea. Nevertheless, conditions were still considered by the engineers to be favorable for the launch, which would take place at 6:00 a.m. Moscow time. Chrétien would later describe the historic occasion in his 2009 book *Rêves d'étoiles* (Dreams of Stars):

> Then suddenly a big light appeared at the horizon. Spectacular and blinding, this light began to go up to the cloud base which it reached in a few seconds, going from this incandescent dot to a magnificent light disc which slowly vanished as Buran ascended into the clouds. Buran lifted off in awful weather, during an inky night. It would return to Baykonur just few hours later and succeed in her first automatic landing. It was also her first flight, without crew. We remained disbelieving, watching the evolution of the weather during those hours. It was evident that our Russian colleagues had taken lots of risks. We learned later that the chances of success were less than 50 percent.
>
> Local television broadcast the rest of the events. Buran began her re-entry into the atmosphere at the right time, and we were anxiously watching pictures from a camera aimed at the base of the clouds on the axis of the landing strip. The ceiling was low, it continued to snow, and there was a strong crosswind blowing. Suddenly Buran broke out of the clouds, some 20 to 30 seconds before landing. We were holding our breath, watching this majestic vehicle beginning its circle before touching down. The landing was magnificent, and Buran stopped majestically amid tremendous applause. Ignoring the risks of getting close too quickly, a group of engineers and technicians ran toward this heroic machine to acclaim it with enthusiasm. Vodka was flowing, even knowing that despite her success it should be her last flight.[4]

RETURNING TO SPACE

Three weeks prior to the launch of the Soyuz TM-7 mission, the two Franco-Soviet crews and a number of Soviet space officials were presented to a group of Western correspondents at a news conference.

In response to a question, Volkov said the flight with Chrétien would create a new space record, as six men would occupy Mir for the first time. He also pointed out that two of those already aboard the station, Vladimir Titov and Musa Manarov, had already been in space for longer than any other humans, having resided on Mir since 21 December the previous year. This easily outdid the 84 days which stood as the longest period that an American astronaut had been in space.

"During our month-long program with the French cosmonaut we will conduct a six-hour spacewalk with him," Volkov told reporters. "He will then go back to Earth with comrades Titov and Manarov on 21 December." This would represent exactly one year since the two long-duration cosmonauts had first entered Mir. The three men were scheduled to return to Earth aboard the Soyuz TM-6 spacecraft. "I will remain on Mir another four months with flight engineer Sergey Krikalyov, who is launching with me and Chrétien."

Also aboard Mir, which had been inhabited continually by rotating teams since February 1987, was Valeriy Polyakov, a space doctor who had arrived on 29 August that same year with the intention of spending longer than a year in space, conducting an extended study of how the human organism adapts to weightlessness. He was scheduled to return home with Volkov and Krikalyov aboard the Soyuz TM-7 spacecraft.

"In western Europe we may have a space station sometime in the late 1990s or afterward," Chrétien said in discussing the upcoming flight, "and the Americans right now don't have a permanently manned space station. We feel it is important to learn all that we can from this opportunity [to visit Mir], and the spacewalk will be a key element for us." Later, Chrétien added a light note to the press conference by revealing he would be taking something special into space with him – music specially recorded for the flight by the British rock group Pink Floyd. "On a flight longer than a couple of days, cosmonauts need recreation," he explained. "Some, like me, like simply sitting at the observation bay and watching the Earth go by, and that experience for me is greatly enhanced by listening to music that I like. I know some of the others on our flight will bring their own electronic keyboards, but I am taking tapes, one of them specially recorded for the flight by Pink Floyd."[5]

Chrétien and Volkov boarding the transfer bus that will carry them out to the launch pad. (Photo: Author's collection)

Lift-off for the Soyuz TM-7 mission. (Photo: Spacefacts.de)

It was 6:50 p.m. Moscow time on 26 November 1988 when the Soyuz TM-7 mission got under way with the successful lift-off of the Soyuz U2 rocket from Baykonur's launch pad 1/5, from where Yuriy Gagarin's historic flight had launched 27 years earlier. In addition to French President François Mitterand, the event was viewed by special guests David Gilmour and Nick Mason from the rock band Pink Floyd.

The Soyuz spacecraft successfully docked with the Mir orbital complex two days later at 8:16 p.m. Moscow time. Once all the checks had been completed, the hatch was opened and Volkov, Krikalyov and Chrétien floated through to be greeted by the resident crewmembers Vladimir Titov, Musa Manarov and Valeriy Polyakov.

Shortly thereafter, Mikhail Gorbachev made a telephone call in which he offered his own greetings and congratulations to all on board Mir and wished the visitors success with their intensive work program. "You have the honor," he began, "of carrying out the second joint Soviet-French space flight. Your mission is of great scientific importance for our countries. It graphically demonstrates the fruitfulness and potential of Soviet-French cooperation and symbolizes the traditional mutual sympathies felt by our peoples."[6]

A FRENCH SPACEWALK

On 9 December at 12:57 p.m. Moscow time, Aleksandr Volkov and Chrétien began their historic spacewalk when they depressurized the multiport docking adapter and clambered outside Mir. Chrétien, who was first out, later said it was "the most fascinating part" of his second space mission. "Probably the most fascinating thing I've ever done in space … just to open the door and get outside … Some people in Star City told me, and they were mission specialists, they were the engineers, so I think these guys did not have flight experience, or maybe parachuting, I don't know. They told me, 'Oh, you will see.

When you open the door, you hesitate. You're really feeling that you will fall, and fall to the Earth, and also during the night you will see, you will stick to the [Mir] space station and you don't want to move because it's so dark.' In fact, none of those things happened.

"It's so fascinating. The door was open. I was looking through at the stars. 'Okay, let's go. Let's go.' When you get out, it's so fantastic, just slowly getting out and see the Earth. You forget about your spacesuit very quickly, so you're really in the impression that you are free-floating with nothing, just swimming."[7]

Soviet television provided live coverage of parts of the spacewalk, which was scheduled to last 4 hours and 20 minutes.

"Then we had to start working. They had asked us to work day and night because we had a lot of different devices to fix. So we started right after we emerged, which was in the dark. But, in fact, there is light enough coming from the albedo of the Earth and also from [lamps on] your helmet. So we started … and the six hours went just like one hour, very, very busy, but really, really fantastic. You don't have time, in fact, to enjoy the fact that you're outside. They keep you very busy, but it's a fantastic experience."[8]

Chrétien during his historic spacewalk outside of the Mir space station. (Photo: Spacefacts.de)

First up, Chrétien installed handrails, then attached the 34-pound Échantillons package to the handrails using springs and hooks. He then hooked up electrical wires leading from the package to Mir's power supply. The two spacewalkers next assembled the 530-pound ERA experimental deployable structure. According to Krikalyov, taking the ERA outside Mir greatly helped to relieve the crowding problems they had experienced inside the station in recent days. After the assembly process Volkov and Chrétien attached a mount to handrails on the frustum linking the multiport docking unit to the small-diameter portion of the work compartment.

After finally resolving issued with cables linking the ERA to a control panel inside Mir, the two men attached the folded ERA structure to a support arm on the platform. The ERA was designed to unfold to make a flat, six-sided structure 1 meter deep by 3.8 meters wide.

From inside Mir, Krikalyov commanded the structure to unfold, but it refused to do more than partially deploy. Knowing this was an important test of construction techniques to be used in weightlessness, and a forerunner for future space station 'building blocks', Volkov and Chrétien kept trying to unfurl the deformable plastic web.

The two spacewalkers fell out of radio contact while scientists at the Kaliningrad control center frantically sought to overcome the problem. When contact was finally restored, the cosmonauts reported that despite their best efforts the structure was still only partially open. They had even tried shaking the structure. "It did not open," Chrétien recalled. "So we had to wait one and a half hours, doing other stuff, because that thing was not working … that delayed our return."[9]

Eventually a frustrated Volkov kicked the ERA. Much to everyone's surprise, this caused the ERA to unfurl properly. The spacewalkers finally made their way back inside Mir after an EVA that had lasted 5 hours and 57 minutes.

WORKING ABOARD MIR

Peter Bond is a lifelong space and astronomy enthusiast and author. In 1990 he wrote an article concerning the Franco-Soviet Aragatz mission, in which he discussed a further ten experiments conducted aboard Mir.

> They included a subscale prototype of a new solar-cell panel, which was deployed inside Mir. Stereoscopic imaging enabled engineers to study the panel's rigidity properties during a series of deployment tests. Long-term studies of microchip degradation by heavy ions also began on the TM-7 mission. Such particles have been identified as the main cause of damage to electronic components in space.
>
> The effects of long-duration space flight on a pilot's ability to command a spacecraft … were evaluated as part of the 'Viminal' experiment. The cosmonauts moved a small handle in response to visual stimuli, so that the validity of their reactions could be ascertained. Mir was also equipped with a radiation dosimeter called 'Circe', which simulated radiation absorption by human tissue. The results were expected to aid protection of future crews from harmful cosmic-rays and solar radiation.[10]

As he worked through his experiments, Chrétien would look forward to his allocated rest period of one day per week – although he would likely have begrudged the mandatory time off of two days per week granted his cosmonaut companions. But while at times the rest of the crew might have experienced difficulties working around Chrétien and his experiments, they would certainly have appreciated the tasty French cuisine he brought along with him, and which he shared with them. It provided a very welcome respite from the normal, bland Soviet menu. There were no less than 23 different meals created by French chefs especially for the Aragatz mission. These included, according to Peter Bond, "pigeon with dates and spices, sauté of vela marengo, boeuf bourguignon, fondue of oxtail, and gruyere and cantal cheeses."[11]

In the latter part of Chrétien's time aboard Mir, he assisted in transferring equipment and experimental results into the Soyuz TM-6 spacecraft, in which he would return home. The three personal seat liners were exchanged, personal effects were loaded, and the checks to confirm the correct center of mass were carried out.

Chrétien tries his hand at making music on Mir using his electric organ. (Photo: Author's collection)

Krikalyov, Chrétien and Volkov aboard Mir, with Chrétien's electric organ floating above them. (Photo: Spacefacts.de)

LANDING AND RECOVERY

On 21 December, Vladimir Titov, Musa Manarov and Jean-Loup Chrétien said goodbye to the three cosmonauts who were to stay aboard Mir, and then entered Soyuz TM-6, closing and sealing the airlock hatch behind them. Titov then undocked their vehicle from the Mir complex and made preparations for re-entry.

After being delayed for three hours by a flaw in the vehicle's automatic control systems the capsule landed safely some 99 miles south-east of Dzhezgazkan in central Kazakhstan. Soviet space experts later said it was possible the ship's computers had overloaded because the crew tried to keep using a section that was normally not utilized during a descent. The three men did not have to wait long for the recovery team to reach them.

As an unnamed observer of the cosmonaut recovery team later recorded in an interview with *Aviatsiya i Kosmonavtika*, the helicopter was heading for the designated recovery area where Titov, Manarov and Chrétien were expected to land. It is worthwhile recording his impressions, which demonstrate the procedures that had to be undertaken following the safe return of a Soviet spacecraft.

> The helicopter was literally fighting its way through a blizzard. Visibility was virtually zero. The helicopter crew and the search and rescue service specialists were calm; they had experienced such conditions during training drills and during the real thing. They had full confidence in the skill of the pilot, Maj. A. Seleverstov. They were certain that he would place the helicopter down alongside the capsule with timeliness and accuracy. But was this flight not dangerous for them? Search and rescue people don't think about themselves.
>
> Powerful search and recovery vehicles equipped with communications transceivers and navigation gear were proceeding along the ground toward the recovery site. Driver-technician 1st class WO [Nikolay] Smetanin was aboard one of these vehicles. In his 12 years of service he has logged more than 100,000 kilometers on the steppe. 'I am on my third revolution of the Earth,' jokes Nikolay Aleksandrovich. During a night recovery in 1983, he was the first to arrive at the landing site of the capsule containing cosmonauts [Vladimir] Lyakhov and [Aleksandr] Aleksandrov, helping them exit from the capsule and driving them to a nearby airfield. He will never forget that experience.
>
> Finally, in a break between snow squalls, we caught sight of the just-landed capsule and a helicopter on the ground alongside it. Lt. Col. N. Vorobyev (Medical Services) loped over to the capsule and peered into the view port. The cosmonauts were smiling and pointing thumbs up: 'A-OK!' Personnel from a search and recovery vehicle which had driven up then opened the exit hatch using a wrench. They pulled out the mission commander first. V. Titov said: 'I've been away for a whole year!' Next came Chrétien. 'Greetings!' he exclaimed and smiled at the recovery people. Manarov was last out; he sat on the edge of the hatch and looked around at the landing site. It was immediately apparent that he was in good health. It was hard to believe that he had been away from the Earth an entire year. It seemed as if he would jump down onto the ground and say, 'Well, why are you standing there with a stretcher?' But procedure must be followed. The cosmonauts were carried into [the medical] tent. Their blood pressure was taken, and other objective indicators were measured. Every three minutes the radio operator transmitted data on the cosmonauts' subjective state of physical well-being. All of the readings were nominal. After a short rest and a 'picture for posterity' taken in front of the tent the cosmonauts climbed unassisted aboard the helicopters, which flew them to Dzhezkazgan; from there they flew by fixed-wing aircraft to Moscow, to Zvyozdnyy Gorodok.[12]

LIFE AFTER SPACE

Following his divorce from Mary-Cathryn, Chrétien remarried on 23 April 1985 to Amy Kristine Jensen from Connecticut in the United States, and a daughter, Lauren, was born November 1989 in France. However, this marriage also ended in divorce. These days his companion is the actress Catherine Alric, with whom he wrote the book *Rêves d'étoiles* in 2009.

In December 1988 Chrétien was promoted to the rank of general. From 1990–1993, as head of the CNES cosmonaut corps, he participated in Buran spacecraft pilot training at the Flight Research Institute of the Ministry of Aviation Industry in Zhukovskiy, south-east of Moscow. He also flew the Tu-154 and MiG-25 aircraft in a role similar to NASA's Shuttle Training Aircraft (STA). He then attended ASCAN (Astronaut Candidate) training at the Johnson Space Center in 1995. He was initially assigned to work on technical issues for the Operations Planning Branch of the Astronaut Office and then served on the crew of STS-86 aboard *Atlantis* in 1997, the seventh shuttle flight to rendezvous and dock with Mir.

When reaching the official French age limit of 60, NASA offered Chrétien the chance to remain in Houston and become a NASA astronaut. He was initially assigned to work for the International Space Station program and later as the deputy director of the ISS Expedition Corporation. At the same time he was assigned to the position of Crew Operation Assistant to George W.S. Abbey, the director of the NASA center. Chrétien later acquired dual U.S.-French citizenship.

Jean-Loup Chrétien in 2012. (Photo: LeParisien.fr)

In September 2000, Chrétien was badly injured while visiting a large hardware store in Webster, Texas, when a 68-pound drill press fell from a shelf more than 10 feet above him. The injuries to his head, neck and shoulders were so severe that a NASA flight surgeon permanently removed him from flight status and he was obliged to retire from the space program in 2001. In that same year his son Olivier died. After leaving NASA his focus shifted to developing a business career in the Houston area and also back in France. At the same time he worked as vice president for research and development at Tietronix, a software company closely linked to the NASA space program. He also became an advisor for space activities to the president of Dassault Aviation and was invited to join the board of Brit Air.

INTERNATIONAL
PROGRAM PATCHES

SOYUZ T-6
(FRANCE)

SOYUZ T-11
(INDIA)

SOYUZ TM-3
(SYRIA)

SOYUZ TM-5
(BULGARIA)

SOYUZ TM-6
(AFGHANISTAN)

SOYUZ TM-7
(FRANCE)

Chrétien currently holds membership of the Academy of Air and Space, the Association of Space Explorers, the Association of European Astronauts, the International Academy of Aeronautics, and the American Institute for Aeronautics and Astronautics. He was awarded the Aeronautics Medal and is a Commander of the Order of the Legion of Honor and also a Knight of the National Order of Merit. Although he is a Hero of the Soviet Union, unlike his Soviet counterparts he was not awarded a second Gold Star after his second mission, as was the custom at that time. Apparently that was thought to be too much honor for a non-Soviet cosmonaut.

In 2010 Chrétien was assigned to the consultative group of the French Government's Grand Emprunt investment/loan program. Today he is still employed as vice president of Tietronix Software.

During 1989–1990, French back-up candidate Michel Tognini supported the Hermes mini-shuttle program in Toulouse, France, before returning to Star City in 1991 to undergo prime crew training for the third Soviet-French space mission named 'Antares'. He was launched for his first space mission on 27 July 1992 aboard Soyuz TM-15. With Anatoliy Solovyev and Sergey Avdeyev, he linked up with Mir and joined the residents Aleksandr Viktorenko and Aleksandr Kaleri. After 14 days spent on a program of joint Soviet-French research he returned to Earth aboard Soyuz TM-14 on 10 August, then returned to France.

In 1995 Tognini underwent ASCAN (Astronaut Candidate) training at the Johnson Space Center in Houston. He was initially assigned to the Operations Planning Branch of the Astronaut Office working on technical issues involving the International Space Station, and was subsequently transferred to the Robotics Branch. He then served as an ISS CapCom in Mission Control. His next assignment was to the flight crew of the STS-93 mission, which was carried out by *Columbia* from 22–27 July 1999, during which his primary tasks were to assist in the deployment of the Chandra X-Ray Observatory and conduct a spacewalk in the event one was needed (it wasn't).[13]

Tognini retired from active astronaut status in May 2003. In January 2005, he became Head of the European Astronaut Centre in Cologne, Germany, then left ESA for retirement on 1 November 2011.

REFERENCES

1. Alma-Ata *Kazkhstanskaya Pravda* newspaper (in Russian), "French Cosmonauts Selected for 1988 Mission with USSR," issue 1 August 1986, pg. 3
2. Neville Kidger, "French Mir Mission Preview," BIS *Spaceflight* magazine, issue November 1988
3. Jean-Loup Chrétien interviewed by Carol Butler for NASA JSC Oral History Project, Houston, Texas, 2 May 2002
4. Jean-Loup Chrétien and Catherine Alric, *Rêves d'étoiles* (Dreams of Stars), Editions Alphée, France, 2009
5. Roman Rollnick article, "Frenchman joins Soviet Cosmonauts aboard Space Station," UPI Archives, 21 November 1988. Online at: *http://www.upi.com/Archives/1988/11/21/Frenchman-joins-Soviet-cosmonauts-aboard-space-station/5002596091600*

6. *Spaceflight News* magazine article, "Soviet-French Mission Launch," edition January 1989, pg. 4

7. Jean-Loup Chrétien interviewed by Carol Butler for NASA JSC Oral History Project, Houston, Texas, 2 May 2002

8. *Ibid*

9. *Ibid*

10. Peter Bond, "The French Connection," *Spaceflight News* magazine, issue November 1990, pp. 14–17

11. *Ibid*

12. S. Skrynnikov, *Aviatsiya I Kosmonavtika* magazine article, published under the heading "People of Duty," (English translation), issue No. 4, April 1989, pp. 21–23

13. Wikipedia entry, *Michel Tognini*, online at: *http://en.wikipedia.org/wiki/Michel_ Tognini*

17

The Interkosmos missions in retrospect

Looking back at the Interkosmos missions, one cannot help but wonder what the main reasons were to fly them in the first place. It was very benevolent of the Soviet Union to provide their socialist satellite states the opportunity to fly a citizen, but clearly it was not intended to be the beginning of a long-term program of joint missions. After all, the participating nations would only be allocated a single mission, lasting just one week. Bulgaria was only offered a second flight because Georgi Ivanov had not been able to reach the Salyut-6 station on Soyuz-33, but that second Soviet-Bulgarian mission was flown on other criteria and conditions than the first one.

It is clear that the flights can be divided in two groups: the first is the series of missions to Salyut-6; the second is the series of missions to Salyut-7 and Mir. And even that first group can be split in two subgroups: one of three missions and one of six missions. To this day, it is unclear what the rationale was behind the order in which the first three were flown, and even why these three were conducted before the others.

In 1990, Sigmund Jähn indicated that Poland had requested to fly the second mission, but he didn't say what that reason was.[1] It could be argued that the Soviets thought it politically undesirable for the Germans to fly first and that this resulted in the order Czechoslovakia – Poland – GDR. It has been suggested that Czechoslovakia was allocated the first mission to guarantee positive propaganda on the eve of the tenth anniversary of the Soviet invasion of Czechoslovakia in 1968 by Warsaw Pact troops, which brought to a sudden end the period of political uncertainty known as the Prague Spring. Demonstrations and violent protests were anticipated, and maybe it was calculated that flying Vladimir Remek at that time would give the Czech people something else to occupy their conversations.

As for Poland, there had been social unrest since 1976 and the situation was deteriorating. The space flight was apparently meant to calm the Polish people a little by demonstrating a positive achievement of the socialist system.

One curious story relating to Polish cosmonaut Miroslaw Hermaszewski featured in the 2009 German television documentary *Fliegerkosmonauten*. In an interview, Hermaszewski claims that he was supposed to fly the first Interkosmos mission but couldn't because of his tonsillitis, "even though I didn't feel any pain". He stated, "I was supposed to fly first. Not the Czech cosmonaut. Unfortunately, it was claimed that I had problems with my tonsils.

© Springer International Publishing Switzerland 2016
C. Burgess, B. Vis, *Interkosmos*, Springer Praxis Books, DOI 10.1007/978-3-319-24163-0_17

Although I felt no pain, my tonsils were removed. Meanwhile, the Czech went into space. The Czech cosmonaut Remek, and I went second, after him." When asked whether he was really ill, Hermaszewski replied, "The doctors said so. After my tonsils were removed, a doctor put them on a table and dissected them. When he cut open the second, he said, 'You were right. They were healthy.'"[2]

Throughout the Interkosmos series there were many indications of the low priority given to these flights – apart from the propaganda activities – as evidenced by the Soyuz-30 flight of Hermaszewski. On the scheduled rest day of the resident crew, the international crew were required to remain in their own spacecraft to perform their experiments so Kovalyonok and Ivanchenkov could have their day of free of work duties. This signified the relatively low importance often given to the Interkosmos missions, in spite of the publicity they generated; otherwise, TsUP could simply have moved or delayed the day of resident crew rest until the international crew had departed.

An explanation for the order of the second subgroup of missions was uncovered by Dutch space enthusiast Maarten Houtman, namely that the international cosmonauts were flown in alphabetical order of their country's names in the Cyrillic alphabet:

Bolgariya	(Bulgaria)
Vengriya	(Hungary)
Vietnam	(Vietnam)
Kuba	(Cuba)
Mongoliya	(Mongolia)
Rumyniya	(Romania)

For the Soviet hierarchy, this had the pleasant advantage that the Romanian cosmonaut would be the last to fly. Relations between the Soviet Union and Romania were not at their best at the time.

As to the second group of flights, there was no specific order. These were decided upon one by one, usually because a joint mission was offered during a state visit.

Salyut-6 was the first Soviet space station that had more than one docking port, enabling extended missions to be flown by crews that would be resupplied by automated cargo ships. Besides that, if crews wanted to fly longer than the operational life of their Soyuz spacecraft, they had to get a new transport ship after a certain time. This replacement ship was brought up by a visiting crew. Because such missions *had* to be flown, the Soviets decided to invite their Interkosmos partners to fly on them.

The first time such a mission was flown was Soyuz-27 with Vladimir Dzhanibekov and Oleg Makarov. Apart from bringing a new Soyuz to the Salyut-6 resident crew of Yuriy Romanenko and Georgiy Grechko, their task was to assess whether the station's systems could support four crewmembers. With the exception of the ill-fated Soyuz 11 trio that visited Salyut-1 in 1971, all Soviet space stations had been occupied by two-man crews. Dzhanibekov and Makarov were launched in Soyuz-27 on 10 January 1978 and returned almost six days later aboard the Soyuz-26 spacecraft. No objections were raised to flying more visiting missions.

Since Soyuz-27 was less than two months old when Aleksey Gubarev and Vladimir Remek lifted off with Soyuz-28 on 2 March 1978, there was no necessity to exchange vehicles at the end of their visit. Romanenko and Grechko returned to Earth only two weeks after Gubarev and Remek, well within the 90-day operational lifetime for Soyuz-27. Therefore, both crews returned in their own ship.

Salyut-6			
UP IN	DOWN IN	CREW	MISSION DURATION
Soyuz-26	**Soyuz-27**	**Romanenko – Grechko**	**Salyut-6 Main Expedition 1**
Soyuz-27	Soyuz-26	Dzhanibekov – Makarov	5 days 22 hrs 58 mins
Soyuz-28	Soyuz-28	Gubarev – Remek	7 days 22 hrs 16 mins

The next Salyut expedition crew of Vladimir Kovalyonok and Aleksandr Ivanchenkov lifted off on 15 June 1978 aboard Soyuz-29, followed only twelve days later by Soyuz-30 carrying Pyotr Klimuk and Polish researcher Miroslaw Hermaszewski. Of course, a swap-over of spacecraft was not necessary at that time. However, by the time Valeriy Bykovskiy and Sigmund Jähn were launched on Soyuz-31 a new ship was needed by Kovalyonok and Ivanchenkov. To facilitate this, the cosmonauts exchanged the seat liners in the two Soyuz vehicles and Bykovskiy and Jähn returned to Earth in Soyuz-29, which had spent 80 days in orbit; ten less than its guaranteed lifetime. Soyuz-31 was left at the station for the residents, who returned on 2 November.

Salyut-6			
UP IN	DOWN IN	CREW	MISSION DURATION
Soyuz-29	**Soyuz-31**	**Kovalyonok – Ivanchenkov**	**Salyut-6 Main Expedition 2**
Soyuz-30	Soyuz-30	Klimuk – Hermaszewski	7 days 22 hrs 2 min
Soyuz-31	Soyuz-29	Bykovskiy – Jähn	7 days 20 hrs 49 min

The Soyuz-33 mishap badly interfered with plans for Salyut-6's third main expedition. Soyuz-32 had been launched on 25 February 1979, followed less than two months later by Soyuz-33. A spacecraft swap was planned, but obviously couldn't occur given Soyuz-33's engine trouble and subsequent emergency landing. Following Rukavishnikov and Ivanov's return, an investigation followed and further manned missions placed on hold. It is unclear when the next mission would have taken place, and a Soyuz exchange may again have been foreseen for that flight. However, as the Soyuz was not cleared in time for its manned role, Soyuz-34 was launched unmanned on 6 June 1979 to provide the station crew with a new spacecraft. Although Soyuz-32 was beyond its operational life, it safely returned to Earth unmanned on 13 June. Lyakhov and Ryumin followed using Soyuz-34 on 19 August.

Salyut-6			
UP IN	DOWN IN	CREW	MISSION DURATION
Soyuz-32	**Soyuz-34**	**Lyakhov – Ryumin**	**Salyut-6 Main Expedition 3**
Soyuz-33	----------	Rukavishnikov – Ivanov (1)	1 day 23 hrs 1 min
Soyuz-34	----------	unmanned (2)	

(1) – Failed to reach the station.
(2) – This replaced Soyuz-32, which in turn came home unmanned.

Leonid Popov and Valeriy Ryumin, the fourth main expedition to the Salyut-6 station, saw two Soyuz vehicles replaced. Soyuz-36 was left by Kubasov and Farkas when they returned in Soyuz-35. Viktor Gorbatko and Pham Tuan returned to Earth in Soyuz-36, leaving Soyuz-37 for Popov and Ryumin. Logic would have dictated that the third visiting crew, with Cuban cosmonaut Arnaldo Tamayo Mendez, would also have changed spacecraft, but they returned in their own Soyuz-38. In all probability, this had to do with the fact that they were followed only 15 days later by Popov and Ryumin in Soyuz-37, which was still within its operational lifetime.

Salyut-6

UP IN	DOWN IN	CREW	MISSION DURATION
Soyuz-35	**Soyuz-37**	**Popov – Ryumin**	**Salyut-6 Main Expedition 4**
Soyuz-36	Soyuz-35	Kubasov – Farkas	7 days 20 hrs 46 min
Soyuz-37	Soyuz-36	Gorbatko – Pham Tuan	7 days 20 hrs 42 min
Soyuz-38	Soyuz-38	Romanenko – Tamayo Mendez	7 days 20 hrs 43 min

The final Salyut-6 main expedition crew of Vladimir Kovalyonok and Viktor Savinykh flew to the station in the new modification of the Soyuz, called Soyuz T. It appears that this mission was mainly flown to fulfill the obligations to Interkosmos member states Mongolia and Romania. Soyuz T-4 was launched just ten days before Soyuz-39 and landed only four days after the second visiting crew departed. The main mission duration of 74 days was well within the limits of the Soyuz's life, therefore exchanging spacecraft hadn't been necessary. On top of that, it is doubtful that the Soviets would have wanted to let either Gurragchaa or Prunariu fly on the new Soyuz, as this would have meant it was necessary for them to learn how to operate two different types of spacecraft.

Salyut-6

UP IN	DOWN IN	CREW	MISSION DURATION
Soyuz T-4	**Soyuz T-4**	**Kovalyonok – Savinykh**	**Salyut-6 Main Expedition 5**
Soyuz-39	Soyuz-39	Dzhanibekov – Gurragchaa	7 days 20 hrs 42 min
Soyuz-40	Soyuz-40	Popov – Prunariu	7 days 20 hrs 43 min

Of course, there was no operational need whatsoever to fly non-Soviet cosmonauts. The rationale for it had everything to do with unleashing publicity and international propaganda. When one looks back at the first series of Interkosmos missions, it is abundantly clear that most of them weren't necessary from an operational standpoint. Sigmund Jähn's Soyuz-31 was the first that *had* to be left behind to supply the resident crew with a new spacecraft for their return to Earth. The same goes for the missions that carried Georgi Ivanov, Bertalan Farkas and Pham Tuan. These flights were all needed to ensure the main expedition crews could fulfill their long-duration missions. But the flights of Remek, Hermaszewski, Tamayo Méndez, Gurragchaa and Prunariu were all manifested without an operational necessity, and only conducted to fly the non-Russian cosmonauts and complete the Interkosmos missions. Given the extraordinary longevity of the Salyut-6 station, it was decided to continue with it. The Soyuz T-4 crew might not have flown the fifth main expedition at all otherwise. Their task was to host the international visitors who arrived in Soyuz-39 and Soyuz-40 and left in their own vehicles.

Jean-Loup Chrétien's first space flight was the last one in which a mission was organized specifically to fly a non-Soviet cosmonaut-researcher. Henceforth, foreigners would only be included on flights that were already planned for a spacecraft swap. By now, the Soviets had reverted to three-man crews and it was a fairly easy operation to allow non-Soviets to fly in the 'spare' seats, where no critical mission tasks had to be conducted. Initially, these seats were offered by the Soviets to countries as a gift during state visits, but after Jean-Loup Chrétien's second flight the Soviets decided to offer such rides either on a fee-paying basis or as part of some broader deal.

Salyut-7

UP IN	DOWN IN	CREW	MISSION DURATION
Soyuz T-6	Soyuz T-6	Dzhanibekov – Ivanchenkov – Chretien	7 days 21 hrs 51 min
Soyuz T-11	Soyuz T-10	Malyshev – Strekalov – Sharma	7 days 21 hrs 40 min

Mir

UP IN	DOWN IN	CREW	MISSION DURATION
Soyuz TM-3	Soyuz TM-2	Viktorenko – Aleksandrov (1) – Faris	7 days 23 hrs 5 min
Soyuz TM-5	Soyuz TM-4	Solovyev – Savinykh – Aleksandrov (2)	9 days 20 hrs 9 min
Soyuz TM-6	Soyuz TM-5	Lyakhov – Polyakov (3) – Mohmand	8 days 20 hrs 26 min
Soyuz TM-7	Soyuz TM-6	Volkov – Krikalyov – Chretien	24 days 18 hrs 7 min

(1)– Russian Aleksandr Aleksandrov.
(2)– Bulgarian Aleksandr Aleksandrov.
(3)– Polyakov remained on board Mir when Lyakhov and Mohmand returned to Earth.

One group was unhappy with this new policy: the Soviet cosmonauts from organizations other than the air force and NPO Energiya. Those nominated by the Institute of Biomedical Problems, the Academy of Sciences, the Ministry of Aviation Industry, and others, saw what they regarded as 'their' seats being sold to foreigners who would never have had a chance to fly if they hadn't come armed with a big bag of American dollars.

Apart from Bulgaria, none of the original Interkosmos countries would see another of their citizens fly in space. There simply never were plans to do that, although it is known that East Germany was in the process of planning the selection of new cosmonauts for possible follow-on missions. After the collapse of the Soviet Union, most of their former satellite states began to look westward and eventually joined the European Space Agency (ESA). The GDR ceased to exist and became part of the Federal Republic of Germany, which was one of the founding members of ESA. The Czech Republic joined in 2008, Romania in 2011, Poland in 2012 and Hungary in 2015. Bulgaria has not (yet) become a member of ESA.

At the time of publication of this book the most recent ESA astronaut selection was that in 2009, which was too early for even the Czech Republic to put forward its own candidates. It is therefore unclear when one of these countries will see their second astronaut (ESA prefers that term to cosmonaut) go into space. With six young astronauts still active, and the limited number of flight opportunities, it will be several years before ESA will need a new group of astronauts. Also, even though ESA officials vehemently deny it, the 1992 and 2009 intakes have clearly established that a candidate's nationality is of crucial importance when it comes to being selected. Because Germany, France and Italy collectively pay the lion's share of the agency's budget, in any selection group there will be reserved opportunities for at least one astronaut candidate from each of these countries. As a result, the smaller countries will only get a chance if there are four or more astronauts selected, and even then, the fact that some countries have their 'own' ESA astronaut is a factor to be considered. Thus the prospects of Czechs, Poles, Hungarians or Romanians being selected for astronaut training any time soon are quite slim.

The Interkosmos program was all about publicity. Nothing less, nothing more. As such, it succeeded and it set several 'firsts' in manned space flight:

- First international spacecraft crew (Gubarev and Remek)
- First non-Soviet and non-American space traveler (Vladimir Remek)
- First Asian cosmonaut (Pham Tuan)
- First black cosmonaut (Arnaldo Tamayo Méndez).

It seems that one thing was closely guarded: all of the original Interkosmos flights which succeeded were just hours short of 8 days' duration. Undoubtedly this was foreseen for the ill-fated Soyuz-33 too. With this scheduled time, no country could claim to have made a significantly longer flight than the others. The small differences were due to the time on the parachute at the end of the mission, because the more wind, the longer it would take before the capsule landed.

When one includes the follow-up international missions, which were not really a part of the Interkosmos program, a few more 'firsts' can be added:

- First Hindu cosmonaut (Rakesh Sharma)
- First Arab/Muslim cosmonaut (Muhammed Faris)
- First EVA by a non-Soviet, non-American cosmonaut (Jean-Loup Chrétien).

Over the years however, these 'firsts' have become less and less significant, especially in comparison to those of the 1960s such as 'first man/woman in space', 'first spacewalk', and the 'first man on the Moon'.

REFERENCES

1. Bert Vis interview with Sigmund Jähn, ASE Planetary Congress, 2–6 July 1990, Groningen, Netherlands
2. Miroslaw Hermaszewski on German TV documentary, *Fliegerkosmonauten*, 2009

18

Philately and the Interkosmos program
by James Reichman

Philately has long played a role in documenting and memorializing significant historic events, notable people, memorable places or things, and their anniversaries. Interkosmos philately not only performed those functions but also went beyond by showing the processes and the traditions of Soviet manned space flight. As a result, the postal productions concerning the Interkosmos programs were exceptional in the minds of cosmic philatelists and even created some rare philatelic gems along the way.

A big part of Interkosmos philately is the stamps that were issued. A number of these are reproduced here. Most readers will be familiar with the concept of stamps that adorn envelopes they receive or send in the mail to indicate that the postal fees were paid, and some may have even taken the time to look at the colorful images on those stamps and contemplate their significance.

Luckily for collectors, philately is much more than just pretty images on stamps. Rather it encompasses a whole range of postal-related products that usually include either an envelope or postcard. When these are franked with stamps and canceled with a postmark they create a philatelic 'cover'. Interest in such a cover is enhanced significantly if the postmark date and post office name have some historic meaning, for example, the launch date of an Interkosmos spacecraft from a post office at or near the Baykonur Cosmodrome. That interest is increased even more if the stamp on the cover directly relates to that same Interkosmos event.

After the dawn of the Cosmic Era (which is considered by the Soviet Union to be the date on which Sputnik-1 was launched, 4 October 1957) and the steady stream of space launches that followed, philatelists around the world started to collect an ever increasing number of philatelic issues and space covers that commemorated those cosmic events. Covers with the actual launch date in the postmark were considered to be launch covers. To this day, covers that were postmarked anywhere in the Soviet Union on that Cosmic Era date continue to be elusive and have become the 'Holy Grail' for many dedicated cosmic philatelists.

As more and more satellites were launched, postal organizations of many nations started issuing space-related stamps featuring those satellites, the rockets that put them in orbit, and the space pioneers who created them. Some even created special postmarks

© Springer International Publishing Switzerland 2016
C. Burgess, B. Vis, *Interkosmos*, Springer Praxis Books, DOI 10.1007/978-3-319-24163-0_18

which featured text and images to commemorate those events. To add even more collector appeal to space covers, some were printed or hand stamped with space-related graphics called 'cachets'.

Top: Moscow postmark dated 13 April 1987 commemorating the 20th anniversary of the Interkosmos Program. Bottom: Soyuz-28 launch cover with a special commemorative Czech postmark. The envelope cachet shows the Cosmonaut Hotel located near the Baykonur Cosmodrome. The cosmonauts stay at that hotel before their space flights and, by tradition, autograph and date their room doors when they set off to the launch site.

A significant milestone was reached in the Cosmic Era when the first man was launched into space and returned safely to Earth. Thus Cosmonaut Yuriy Gagarin became an instant international celebrity – a name known by all space enthusiasts. Stamps bearing his image were issued, monuments were raised in his honor, and even ships were named after him.[1]

This event produced a significant boost for interest in cosmic philately – new stamps, special postmarks, and new space-related covers. In addition, with manned space flight there was not only a desire to obtain a launch cover but now also to obtain a landing cover in order to properly document the beginning and ending dates and locations of a space flight. Such interest in manned space flight helped push the collecting of space-related covers, stamps, and postmarks to new heights, thereby making it the most popular philatelic collectible area in the world.

As the Soviet Union racked up an ever increasing list of space 'firsts' (first satellite, first man in space, first space probe to the Moon, etc.), each was commemorated with philatelic collectables not only in the Soviet Union but also by the postal organizations of their closest socialist allies. Besides helping the Soviets to celebrate their space achievements, these fraternal socialist countries also had interests in participating in those space programs. Eventually a broad program of space cooperation was decided on, and on 13 April 1967 an agreement was signed that created a program called Interkosmos.[2] In order to commemorate the 20th anniversary of that auspicious program's beginning, a special Interkosmos philatelic postmark was introduced at a Moscow post office on 13 April 1987.

Every year after Gagarin's space flight, the Soviet Union conducted an ever more complex series of manned space flights in orbit around the Earth. One such operation in January 1969 involved the docking of the Soyuz-4 and Soyuz-5 spacecraft. Not only did this docking lead to an historic transfer of two crew members from one vehicle to the other by 'spacewalking', it also established a new philatelic tradition of sending mail up to the crewmen in space. This type of out-of-this-world mail service, and its counterpart of creating and posting mail while in space for delivery after the next spacecraft returned to Earth, resulted in a new philatelic tradition related to manned space flight called 'Cosmic Post'.

Moving on, the Soviet Union began launching space stations which could be occupied by manned crews that were flown there and back to Earth on Soyuz spacecraft. This also led to another space flight milestone which cosmic philatelists liked to commemorate, which is the docking of the Soyuz spacecraft to such a space station. Thus each flight to a Salyut or later the Mir space station, had corresponding covers to commemorate that launch, docking, and eventual return to Earth of that crew.

After the initial euphoria of manned space flight, when the Soviet Union issued multiple stamps to commemorate each flight and its crew, the Soviet postal organization settled down and typically issued just one commemorative stamp for each space flight. In addition, only after a spacecraft was successfully recovered back on Earth did the Soviets issue stamps that included portraits of the cosmonauts involved. At least they *eventually* issued stamps with the cosmonaut portraits. This helped to add extra collector interest in Soviet manned space flight stamps, because the Americans never issued stamps with portraits of their astronauts. This was part of the American postal service policy of not including portraits of people on their stamps until at least 10 years after a person was dead.[3]

Top: Special Soviet 'Cosmic Post' postmark for canceling Interkosmos covers created onboard a Salyut space station. Bottom: Soviet cover commemorating the docking ('СТЫКОВКА') of the USSR-Hungary Interkosmos spacecraft Soyuz-36 to the Salyut-6 space station.

The Soviet policy did at least insist on waiting until after the space flight was over before it would print stamps bearing cosmonaut portraits. This was a prudent decision because of the ever-present possibility that the cosmonaut initially chosen to make the space flight might be dropped from the crew, perhaps due to illness or injury, and be replaced by a back-up.

It was also prudent for stamp designers not to include other flight details or images for events that were planned for the flight but not yet achieved, due to the uncertainties in the execution phase of those mission plans. Unfortunately, despite all of their experience in preparing stamps to commemorate space launches, even the Soviet postal service erred in producing a 15-kopeck stamp to commemorate the activities of the Bulgarian cosmonaut onboard the Salyut-6. These activities didn't occur because the Soyuz-33 spacecraft was unable to reach the space station. Although the Soviets did not release these stamps, some were unofficially saved from destruction and made it into collector hands, thus creating the first of a few rare, philatelic gems related to the Interkosmos program.[4]

Each of the Interkosmos countries participating in the manned space flights approached the philatelic commemoration of their cosmonauts' exploits in a different manner. All but one of those countries issued commemorative stamps honoring their space flight. They also created and used special postmarks to cancel those stamps before, during, or after the actual mission events. Also popular among those issues were stamps and postmarks released on the various post-flight event dates (like the return of a cosmonaut to his home country) and the first-year anniversary of the flight.

Left: An un-issued Soviet stamp for the Soyuz-33 USSR-Bulgarian space flight. Right: Polish special postmark commemorating Cosmonaut Hermaszewski's homecoming to the Silesia region of Poland.

Given their relative inexperience at issuing stamps on or near the date of a manned space flight, it is a wonder that so few of these non-Soviet, philatelic items suffered from mistakes. Czechoslovakia, for example, avoided problems by issuing commemorative stamps on their cosmonaut's launch date that were just previously-issued stamps over-printed with the simple phrase 'Joint Flight, USSR-CSSR' without any names or a date. They waited until the first-year anniversary of that launch to release stamps with any real details and images related to their Soyuz-28 flight.

The Polish postal organization wasn't so lucky. Despite the fact that Hermaszewski was always the primary choice to fly the joint flight,[5] the Polish stamp designers were some-how misled into believing that Zenon Jankowski would be the first Pole into space. Because of this, they created and printed up two different stamp designs with his portrait. Luckily for them, the plan was to release these stamps only after the spacecraft was in orbit. Sometime before the launch, they learned of their mistake and were forced to redesign and reprint the stamps with Hermaszewski's image. One published account about this affair says that they did not learn of this situation until the day before liftoff![6] If true, this would have caused a monumental effort to hastily change both of these stamps and print sufficient copies to be issued on the launch date.

Precisely when the Polish philatelic agency really knew Hermaszewski was going to fly and Jankowski wasn't, is not known. The story that it was on the day prior to launch seems farfetched and probably exaggerated to sensationalize the events. Regardless of when the error was discovered, the entire printing of stamps with the wrong cosmonaut was a wasted effort and the stamps were never issued. Despite this some escaped the shredding machine and became philatelic gems in the hands of a few lucky collectors.[7]

Left: Correct Cosmonaut – Hermaszewski's portrait on an issued Polish stamp. Right: Wrong Cosmonaut – Jankowski's portrait on an un-issued Polish stamp.

The Indian postal organization tried to avoid the problem of a last minute change of crew by creating a stamp that included portraits of all of the Soyuz T-11 crewmembers in training, just in case the back-up crew was flown instead of the prime crew. After those stamps were produced, but before they were issued, the Soviets were obliged to replace one of their men after he became ill. This caused the makeup of the crews to be different than that shown in the Indian stamp and therefore all copies of it were supposedly destroyed. Once again a rare Interkosmos philatelic gem was created when at least one unofficial copy of that stamp made it into a collector's hands.[8]

Interestingly, the Syrian postal organization threw caution to the wind and released three commemorative stamps on the day that their cosmonaut was launched. Boldly disregarding the possibilities of space flight mission changes, one stamp commemorated that launch, one the upcoming docking with Mir, and one commemorated the landing of the spacecraft, still more than a week away when the stamp was released. The scheduled dates of those events were included in the stamp designs. Lucky for them, there were no glitches in the schedule and all events turned out exactly on the dates printed on those stamps!

The Soviet stamp designers, for their part, made their stamp designs generic enough to avoid many of these last-minute change pitfalls. Dates on their stamps, for example, were

The un-issued Indian stamp showing Soyuz T-11 flight crews before the last-minute crewmember change.

limited to just the year. In addition, only about half of their issues contained images of the Interkosmos cosmonauts. The faces on all of those images are so small (under 2 mm) that many appear generic, so that few people could say which of the two cosmonauts (prime or back-up) those stamp images represented. Whether or not this was a conscious decision by the designers, so that these stamps could represent either cosmonaut no matter who flew, is not known.

A second Soviet design decision, at least for most of the joint Interkosmos flights, was to create three stamps for each flight instead of just one like they had been doing for years for their own manned missions.[9] The problem was to decide what particular symbolism should be included in those issues that would not make each space flight's set of stamps look like a repeat of the others. Each set of three stamps was to show a chronological sequence whilst also including a different group of snapshots with insights into what steps a typical manned space flight included. What they ended up creating (once all the Interkosmos stamps were completed and then reordered into a training and flight event timeline) was a set of visual images that showed the overall process that went into conducting a joint manned flight. To add enough variety to these sequenced sets, the images included not only the required events (e.g., launch, docking, and landing) but also the traditions interwoven around those events.

Tradition, as well as a good deal of superstition,[5] have always played a part in the Soviet manned space flight process. Some traditions, like the offering of bread and salt to visiting cosmonauts, is deeply rooted in Russian culture. More specifically space-related traditions have evolved since Yuriy Gagarin's epic, first manned space flight in 1961. Some of these traditions are taken from the steps and actions Gagarin took on that first, highly successful flight. Following this ritual, every cosmonaut wants to take those same steps and perform those same actions. This is done partly to respect Gagarin, but it is also superstitious – like 'knocking on wood' – to ensure that their own flight will be just as successful.

Gagarin's public status after his space flight was that of a superhero. After his death in 1968 his status was promoted (in the words of one analyst) to that of a demi-god.[5] His statue at the cosmonaut training complex is but one of many such monuments to him found across the USSR. The training center monument, in particular, is the one that the cosmonauts and their families, by tradition, place flowers on in order to honor him and to express their hope for success (if prior to their space flight) or their thanks for having had a safe flight (if afterwards). "He called us all into space," said American astronaut Neil Armstrong, who visited this training center and wrote those profound words in the guest book at the Gagarin office-museum.[10] In some respects, those words must resonate strongly with his fellow cosmonauts who believe that Gagarin will also help protect them while they are in orbit if they but follow his example.

East German cosmonaut Sigmund Jähn and his family place flowers at Gagarin's cosmonaut training center statue.

That is not to say that all of the traditions that Gagarin started were necessarily good ones. Take, for example, the tradition that he unwittingly started by urinating on the transport bus tire after it stopped during the ride from the cosmonaut suiting-up area to the launch site. It was undoubtedly an urgent necessity as he contemplated enduring all of the press questioning and official ceremonies at the base of the rocket and then the long period of being strapped in the capsule to await the launch. Despite the uncouth aspects of this event, it was picked up as just another ritual that cosmonauts must do in order to follow in Gagarin's steps to success.[11]

Thankfully that particular manned space flight tradition was not featured in the images of the stamps issued to commemorate the Interkosmos flights. But some other traditions were, and these are included in the color plate entitled 'Soviet Manned Space Flight Traditions'.

Interkosmos Launch Covers

Soyuz-40, USSR-
Romanian Launch
on 14 May 1981

Soyuz T-6, USSR-
French launch on
24 June 1982

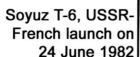

Soyuz-33, USSR-
Bulgarian launch on
10 April 1979

Soyuz-28, USSR-
Czechoslovakia launch
on 2 March 1978
with Cosmonaut
Remek autograph

Interkosmos Spaceflight Patches

Soyuz-31 Spaceflight

Soyuz-36 Spaceflight

Soyuz-37 Spaceflight

Soyuz-28 Spaceflight

Soviet Interkosmos Sheet with ring of flight patches

Soyuz TM-6 Spaceflight

<--- Soyuz T-11 Spaceflight
Patch in Envelope Cachet

Manned Spaceflight Process

Testing trainee's physical
ability to withstand launch

Soyuz and Salyut
simulator training

Mating Soyuz craft
to launch rocket

Transporting rocket to
launch pad

Rocket launch

Soyuz docking to
Salyut-6

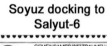

Performing on-orbit
scientific tests

Re-entry and
parachute landing

Post-landing checks

Soviet Manned Spaceflight Traditions

Laying flowers at
Gagarin's monument

Waving to crowd from
launch gantry

Writing on re-entry
capsule after landing

Canceling flight covers onboard
Salyut using special Cosmic Post
postmarks

Televideo conference from space
with dignitaries and newsmen

Presenting bread and
salt to visiting
cosmonauts

Getting red-carpet
reception at Moscow
airport after a
successful spaceflight

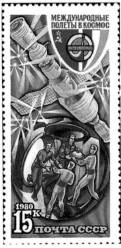

The early Interkosmos manned space flights also created opportunities to start their own traditions. One of these traditions was for the postal service of the guest country to create a special 'Cosmic Post' postmark that the Soviet postal service would allow to be used while canceling Soviet postal issues. These typically had text in both Russian and the language of the guest cosmonaut, and were intended for use onboard the space station to cancel covers commemorating the joint space flight.

Czech postmark for Soyuz-28.

Covers created under these in-orbit conditions are part of a class of philatelic items called 'flown covers' and they are highly sought after by cosmic philatelists. However, because of size and weight limitations in the Soyuz spacecraft, the numbers of these covers that could be created were limited. Because of those restrictions, these flown space covers are some of the philatelic gems that the Interkosmos program helped to create.

Unfortunately, the Soviet postal organization typically allowed for duplicate canceling devices to be made, so that ground-equivalent covers could be created to supplement those created in space. This was done to help meet the expected worldwide demand by collectors for such items. Most of the time these duplicate cancels had some small, barely perceptible modifications in their design which enabled them be distinguished from those original ones used in orbit. However, many casual collectors don't have access to the secret-decoder-ring information that allows them to tell the difference between the philatelic gems and their not-so-valuable copies. For these reasons, collectors can be easily duped into thinking a cover was flown in space when actually it wasn't.

Poland postmark for Soyuz-30.

Mongolia postmark for Soyuz-39.

A second Interkosmos-type tradition is concerned with crew patches. Perhaps these were not a new tradition for manned space flight, but one aspect was fairly new. That was for the stamps that commemorated those international flights to include images of the crew patches in their stamp designs. This began when all of the Soviet stamps issued to commemorate the joint flight with the Americans, known by the Americans as the 'Apollo-Soyuz Test Project' and in the Soviet Union as the 'Soyuz-Apollo Test Project', featured an image of the jointly designed crew patch. The inclusion of flight-patch images in stamp designs was new for the Soviets, because none had been previously included in a stamp.

The 1983 Soviet souvenir sheet included flight patch images for Apollo-Soyuz and the first ten Interkosmos space flights.

A Czech stamp showing the Soyuz-28 re-entry capsule in the Prague Military Museum in the city of Kbely.

The Interkosmos efforts already had an overall program logo (a circle with a star sweeping upward though its center and the word 'Interkosmos') which could have been used as a flight patch, but the interagency groups working on the publicity for these missions wanted to have distinctive crew patch designs which would uniquely identify the two participating countries associated with each space flight.[12] In the end, two flight patches were used for each flight. There was a generic one bearing the Interkosmos logo and a second one unique to each joint space flight. Typically, the Interkosmos flight patch was sewn on the right chest area of each cosmonaut's spacesuit, while the unique crew patch was sewn onto the right upper-arm area.

Of the 35 stamps the Soviets created to commemorate the joint Interkosmos flights, none included a flight patch in their design. However, 34 of the stamps issued by the non-Soviet participating countries included images of their unique crew patch. It wasn't until 1983 that the Soviet Union, as part of a philatelic commemoration of Cosmonautics Day for that year, issued a stamp that included a ring of eleven of these joint crew patch designs encircling the Interkosmos logo.

One final and new space flight tradition began with the design of an Interkosmos-related stamp issue. The Soviets decided to donate to each Interkosmos country the Soyuz re-entry capsule in which their cosmonaut had returned to Earth, for display in a local museum. The only Interkosmos stamp to document this situation was a Czech stamp issued on 7 February 1985.[13] A few examples of the other re-entry capsules that are found in the museums of the Interkosmos participants include Soyuz-30 at the Military Technology Museum in Warsaw, Poland; Soyuz-29 in the Military History Museum in Dresden, Germany; Soyuz T-6 in the *Musée de l'Air* near Paris, France; and Soyuz T-10 in the Nehru Planetarium in New Delhi, India.

Although philately has assisted in determining the final resting place of at least one of the Interkosmos space capsules, it can hardly be considered the first choice for historians trying to research and understand what these joint manned space flights were all about. This is due, in part, to the fact that no comprehensive report has been published which adequately details all the Interkosmos philatelic issues, their related covers, and the backgrounds behind each. Nevertheless, Interkosmos philately does a good job of helping people to visualize the steps that were necessary to actually perform such space flights and the traditions related to them.

REFERENCES

1. Reichman, James & Bartos, Alec, *Philatelic Study Report 2014-1, Soviet & Russian Space-Support Ship Covers and Cachets*, privately printed, 30 November 2014.
2. Petrovich, G. W., *The Soviet Encyclopedia of Space Flight*, Mir Publishers, 1969, pp. 199–200.
3. This US policy changed in 2011. See Associated Press article "Living people to appear on US stamps", *USA Today*, 26 September 2011.
4. Scan of un-issued Soyuz 33 15-kopeck stamp from collection of Leo Malz.
5. Burgess, Colin email, Subject: "Bert Vis on Philatelic Piece", dated 19 October 2014. Includes forwarded comments by Soviet space analyst Bert Vis on various Interkosmos subjects. One of these involves private discussions between Vis and Hermaszewski in

which the cosmonaut says that he was always to be the prime Polish cosmonaut on the joint space flight.

6. Hooper, Gordon R., *The Soviet Cosmonaut Team*, GRH Publications, 1986, p. 186.
7. Scan of un-issued Soyuz-30, wrong-cosmonaut stamp from the collection of Leo Malz.
8. Un-issued Indian stamp image from the collection of Walter Hopferwieser.
9. In part, this decision was for political propaganda. See Dugdale, Jeff, "Soviet Designers Promoted Space Achievement in Propaganda Issues", *Orbit*, Issue #101, March, 2014, pp. 12–15.
10. Riabchikov, Evgeny, *Russians in Space*, Doubleday & Co., 1971, p. 277.
11. See: http://www.blastr.com/2012/08/why_do_russian_cosmonauts.php
12. See: http://www.spacepatches.nl/salyut/s28.html
13. See: http://en.wikipedia.org/wiki/Prague_Aviation_Museum,_Kbely

Appendix A

Interkosmos unmanned satellite program

MISSION	SATELLITE TYPE	LAUNCH DATE	LAUNCH VEHICLE	LAUNCH BASE
Kosmos-261	DS-U2-GK	19.12.1968	Kosmos-2	Plesetsk
Interkosmos-1	DS-U3-IK-1	14.10.1969	Kosmos-2	Kapustin Yar
Interkosmos-2	DS-U1-IK-1	25.12.1969	Kosmos-2	Kapustin Yar
Kosmos-348	DS-U2-GK	13.06.1970	Kosmos-2	Plesetsk
Interkosmos-3	DS-U2-IK-1	07.08.1970	Kosmos-2	Kapsustin Yar
Interkosmos-4	DS-U3-IK-2	14.10.1970	Kosmos-2	Kapustin Yar
Vertikal-1	-	28.11.1970	V-5V	Kapustin Yar
Vertikal-2	-	20.08.1971	V-5B	Kapustin Yar
Interkosmos-5	DS-US-IK-2	02.12.1971	Kosmos-2	Kapustin Yar
Interkosmos- 6	Energuia No. 1	07.04.1972	Voskhod	Baikonur
Interkosmos-7	DS-U3-IK-3	30.06.1972	Kosmos-2	Kapustin Yar
Interkosmos 8	DS-U1-IK-2	30.11.1972	Kosmos-2	Plesetsk
Interkosmos-9 (Kopernik-500)	DS-U2-IK-8	19.04.1973	Kosmos-2	Kapustin Yar
Interkosmos-10	DS-U2-IK-3	30.10.1973	Kosmos-3M	Plesetsk
Interkosmos-11	DS-U3-IK-4	17.05.1974	Kosmos-3M	Kapustin Yark
Interkosmos-12	DS-U2-IK-4	31.10.1974	Kosmos-3M	Plesetsk
Interkosmos-13	DS-U2-IK-5	27.03.1975	Kosmos-3M	Plesetsk
-	DS-U3-IK	03.06.1975	Kosmos-3M	Kapustin Yar
Vertikal-3	-	03.09.1975	V-5V	Kapustin Yar
Interkosmos-14	DS-U2-IK-6	11.12.1975	Kosmos-3M	Plesetsk
Interkosmos-15	AUOS-3-T-IK	19.06.1976	Kosmos-3M	Plesetsk
Interkosmos-16	DS-U3-IK-5	27.07.1976	Kosmos-3M	Kapustin Yar
Vetikal-4	-	14.10.1976	V-3A	Kapustin Yar
Vetikal-5	-	30.08.1977	V-3A	Kapustin Yar
Interkosmos-17	AUOS-3-R-E-IK	24.09.1977	Kosmos-3M	Plesetsk
Vetikal-6	-	25.10.1977	V-3A	Kapustin Yar
Interkosmos-18 (Magion)	AUOS-3-M-IK	24.10.1978	Kosmos-3M	Plesetsk
Vertikal-7	-	03.11.1978	V-3A	Kapustin Yar
Interkosmos-19	AUOS-3-I-IK	27.02.1979	Kosmos-3M	Plesetsk

(continued)

© Springer International Publishing Switzerland 2016
C. Burgess, B. Vis, *Interkosmos*, Springer Praxis Books, DOI 10.1007/978-3-319-24163-0

(continued)

MISSION	SATELLITE TYPE	LAUNCH DATE	LAUNCH VEHICLE	LAUNCH BASE
Vertikal-8	-	29.09.1979	V-3A	Kapustin Yar
Interkosmos-20	AUOS-3-R-P-IK	01.11.1979	Kosmos-3M	Plesetsk
Interkosmos-21	AUOS-3-R-P-IK	06.02.1981	Kosmos-3M	Plesetsk
Interkosmos-22 (Bolgaria-1300)	IK-B-1300	07.08.1981	Vostok-2M	Plesetsk
Vertikal-9	-	28.08.1981	V-3A	Kapustin Yar
Oreol-3	AUOS-3-M-A-IK	21.09.1981	Tsiklone-3	Plesetsk
Vertikal-10	-	21.12.1981	V-3A	Kapustin Yar
Vertikal-11	-	20.10.1983	V-3A	Kapustin Yar
Interkosmos-23	SO-M	26.04.1985	Molnia-M	Baikonur
Interkosmos-24 (Magion-2)	AUOS-3-AV-IK	28.09.1989	Tsiklone-3	Plesetsk
Interkosmos-25 (Magion-3)	AUOS-3-AP-IK	18.12.1991	Tsiklone-3	Plesetsk
Koronass-1	AUOS-SM-KI-IK	02.03.1994	Tsiklone-3	Plesetsk

Appendix B

Interkosmos crewing

SOYUZ MISSION	GUEST NATION	PRIME CREW	BACKUP CREW	LAUNCH DATE	SPACE STATION
Soyuz-28	Czechoslovakia	Aleksey Gubarev Vladimír Remek	Nikolay Rukavishnikov[1] Oldřich Pelčák	03.02.1978	Salyut-6
Soyuz-30	Poland	Pyotr Klimuk Mirosław Hermaszewski	Valeriy Kubasov Zenon Jankowski	27.06.1978	Salyut-6
Soyuz-31	East Germany	Valeriy Bykovskiy Sigmund Jähn	Viktor Gorbatko Eberhard Köllner	26.08.1978	Salyut-6
Soyuz-33	Bulgaria	Nikolay Rukavishnikov Georgi Ivanov	Yuriy Romanenko Aleksandr Aleksandrov	10.04.1979	Salyut-6[2]
Soyuz-36	Hungary	Valeriy Kubasov Bertalan Farkas	Vladimir Dzhanibekov Béla Magyari	26.05.1980	Salyut-6
Soyuz-37	Vietnam	Viktor Gorbatko Pham Tuân	Valeriy Bykovskiy Bùi Thanh Liêm	23.07.1980	Salyut-6
Soyuz-38	Cuba	Yuriy Romanenko Arnaldo Tamayo Méndez	Yevgeniy Khrunov José Armando López Falcón	18.09.1980	Salyut-6
Soyuz-39	Mongolia	Vladimir Dzhanibekov Zhugderdemidiyn Gurragchaa	Vladimir Lyakhov Maidarzhavyn Ganzorig	23.03.1981	Salyut-6
Soyuz-40	Romania	Leonid Popov[3] Dumitru Prunariu	Yuriy Romanenko Dumitru Dediu	14.05.1981	Salyut-6
Soyuz T-6	France	Vladimir Dzhanibekov[4] Aleksandr Ivanchenkov Jean-Loup Chrétien	Leonid Kizim Vladimir Solovyov Patrick Baudry	24.06.1982	Salyut-7
Soyuz T-11	India	Yuriy Malyshev Gennadiy Strekalov[5] Rakesh Sharma	Anatoliy Berezovoy Georgiy Grechko Ravish Malhotra	02.04.1984	Salyut-7

(continued)

© Springer International Publishing Switzerland 2016
C. Burgess, B. Vis, *Interkosmos*, Springer Praxis Books, DOI 10.1007/978-3-319-24163-0

(continued)

SOYUZ MISSION	GUEST NATION	PRIME CREW	BACKUP CREW	LAUNCH DATE	SPACE STATION
Soyuz TM-3	Syria	Aleksandr Viktorenko Aleksandr Aleksandrov[6] Muhammad Faris	Anatoliy Solovyov Viktor Savinykh Munir Habib	22.07.1987	Mir
Soyuz TM-5	Bulgaria	Anatoliy Solovyev Viktor Savinykh Aleksandr Aleksandrov	Vladimir Lyakhov Aleksandr Serebrov[7] Krasimir Stoyanov	O6.07.1988	Mir
Soyuz TM-6	Afghanistan	Vladimir Lyakhov Valeriy Polyakov Abdul Ahad Mohmand	Anatoliy Berezovoy German Arzamazov Mohammad Dauran	29.08.1988	Mir
Soyuz TM-7	France	Aleksandr Volkov Sergey Krikalyov[8] Jean–Loup Chrétien	Aleksandr Viktorenko Aleksandr Serebrov Michel Tognini	26.11.1988	Mir

1. Replaced original backup commander Yuriy Isaulov
2. Docking failure
3. Replaced original commander Yevgeniy Khrunov
4. Replaced original commander Yuriy Malyshev
5. Replaced original flight engineer Nikolay Rukavishnikov
6. Soviet cosmonaut: not to be confused with Bulgarian of the same name
7. Replaced original crewmember Andrey Zaytsev
8. Replaced original flight engineer Aleksandr Kaleriy

About the authors

Colin Burgess Australian author Colin Burgess grew up in Sydney's southern suburbs where he and his wife Patricia still live. They have two grown sons, two grandsons and a granddaughter.

His working life began in the wages department of a major Sydney afternoon newspaper, where he first picked up the writing bug, and later as a sales representative for a precious metals company. He subsequently joined Qantas Airways as a passenger handling agent in 1970 and two years later transferred to the airline's cabin crew. He would retire from Qantas as an onboard Flight Service Director/Customer Service Manager in 2002, after 32 years' service.

During that period several of his books were published on the Australian prisoner-of-war experience, as well as the first of his biographical books on space explorers such as Australian payload specialist Dr. Paul Scully-Power, and *Challenger* teacher Christa McAuliffe. He has also written extensively on space flight subjects for astronomy and space-related magazines in Australia, the United Kingdom and the Unites States.

In 2003 the University of Nebraska Press appointed him Series Editor for their ongoing *Outward Odyssey* series of books detailing the history of space exploration, and he was involved in co-writing three of these volumes. His first Springer-Praxis book, *NASA's Scientist-Astronauts*, co-authored with British-based space historian David J. Shayler, was released in 2007. *Interkosmos* will be his 10th title with Springer-Praxis, for whom he is currently researching and writing a series of books for their *Pioneers in Early Spaceflight Series* on the two suborbital and four orbital space missions in NASA's manned Mercury program.

Bert Vis Bert Vis was born in Voorburg, The Netherlands where he still lives. He has been working for the local and later regional fire service for over 35 years. After almost 34 years in the operational service he became a policy advisor on infrastructural subjects.

He first became interested in the history of manned space flight when Apollo 8 was launched in December 1968, carrying the first human crew to travel to the moon and back. From there his interest grew and in 1975 he visited the NASA Johnson Space Center in Houston for the first time.

In later years he began corresponding with Soviet/Russian cosmonauts and soon found his interest was concentrated on unflown cosmonauts, who, for one reason or another, had never made it into orbit. This active correspondence eventually resulted in an invitation to visit the Yuriy Gagarin Cosmonaut Training Center in Star City, near Moscow, in 1991. Since then, he has traveled to the training center over 20 times, attending official functions

© Springer International Publishing Switzerland 2016
C. Burgess, B. Vis, *Interkosmos*, Springer Praxis Books, DOI 10.1007/978-3-319-24163-0

and conducting extensive interviews with cosmonauts and other space flight officials – many of which were utilized in putting this book together. As well, he has attended most of the annual congress meetings of the international Association of Space Explorers, held in different countries each year. His visits, research and interviews eventually resulted in the first, and so far only, English language book on the history of the cosmonaut training center, *Russia's Cosmonauts: Inside the Yuri Gagarin Training Center*, co-authored with fellow researchers Rex Hall and David Shayler, and published in 2005.

In addition, together with fellow Soviet space historian Bart Hendrickx, he co-wrote the comprehensive study *Energiya-Buran, The Soviet Space Shuttle* (released in 2007). Both books were published by Springer-Praxis.

James Reichman While living in Florida in the mid-1950s, James Reichman became fascinated by rocket launches at nearby Cape Canaveral. He began stamp collecting at the age of 9, a hobby which centered on worldwide issues related to space, and later narrowed to just Soviet space flight.

After graduating from college he was commissioned an officer in the U.S. Air Force and assigned to the Space Defense Center in Colorado Springs, Colorado. While there, he served in positions involving orbital analysis, space surveillance, space object identification and on-orbit satellite control. His 20 years with the USAF included such key positions as Chief of the Satellite Classification and Mission Identification Branch, Chief of the Space Requirements and Test Branch, and Deputy Director of Space Division's Intelligence Directorate.

Retiring from the USAF in 1987, he went to work as a Senior Systems Analyst for the Science Applications and International Corporation. After 35 years of working in spaceflight-related positions he fully retired in 2002.

In 1975 he began writing and publishing articles about his Soviet space flight collecting interests, and subsequently had articles published in American and worldwide philatelic journals including *The Astrophile, Linn's Stamp News*, *Orbit* (journal of the Astro Space Stamp Society) and *Ad Astra* (journal of the Italian Astrophilately Society). He has also performed detailed studies on various aspects of Soviet cosmic philately, and to date has published 19 reports based on those analyses.

Index

A

Abbey, George W.S., 281
Aerospatiale, 201, 269
Aksyonov, Vladimir, 116, 190
Alashki, Minko, 79
al-Assad, President Bashar, 235
Aleksandrov, Aleksandr (Bulgarian cosmonaut),
 225, 248
Aleksandrov, Aleksandr (Soviet cosmonaut),
 83, 245
Aleksandrov, Blagovesta, 240
Aleksandrov, Panayot, 80, 238
Aleksandrov, Panayot (son), 240
Aleksandrov, Plamen, 238, 239
Aleksandrov, Radoslav, 240
Alric, Catherine, 281
al-Zurf, Gen. Abdullah, 229
American Institute for Aeronautics and
 Astronautics, 283
Anadolu (news agency), 235
Andersen, Hans Christian, 70
Apollo-Soyuz Test Program (ASTP),
 90, 98
Arabi, Kamal, 224
Archenhold Observatory, Berlin, 73
Armstrong, Neil, 298
Ars Polona (philatelic agency), 52
Arzamazov, German, 255
Association of European Astronauts, 283
Association of Space Explorers (ASE), 53, 88,
 180, 283
Asteroid No. 2552, 31
Asteroid No. 6149, 33
Atkov, Oleg, 216–218
Avdeyev, Sergey, 283
Aviatstya i Kosmonavtika (magazine), 280

B

Baldangiin, Zhugderdemidiyn, 143
Balebanov, Vyacheslav, 199

Bandera, Stepan, 35
Baranski, Stanislaw, 45
Basescu, Traian, 179
Batista, Fugencio, 127
Baudry, Chantal, 188
Baudry, Claude, 189, 203
Baudry, Liliane, 188
Baudry, Mélodie, 189
Baudry, Nicole, 188
Baudry, Odette, 188
Baudry, Patrick, 170, 187, 191–193, 200–203
Baudry, Philippe, 188
Baudry, Roger, 188
Baudry, Stéphanie, 203
Bella, Ben, 235
Beregovoy, Georgiy, 63, 116, 227
Berezovoy, Anatoliy, 101, 193, 200, 212, 255
Berger, Rolf, 60–62, 74, 75
Betmunkh, Dashzeveiin, 143
Bialon, Pavol, 15
Blagov, Viktor, 230, 232
Blaha, Antonin, 29
Blaha, John, 29
Boleski, Marian, 38
Bloszczynski, Romuald, 41, 47
Boback, Heinz, 60
Bond, Peter, 278
Bonev, Prof. Boris, 238, 248
Borodin, Aleksandr, 255, 256
Brasov, Romania, 159
Brezhnev, President Leonid, 1, 25, 106, 123, 136,
 137, 175, 186, 205, 206
Brno, Moravia, 13, 28, 31
Bucharest Polytechnic Institute, 159
Buczko, Imre, 95, 96
Budapest Technical College, 107
Bugala, Andrzej, 38–41, 43
Bulgarian Academy of Sciences, 90, 238, 240,
 247–250
Bykovskiy, Valeriy, 64, 65, 72, 79, 101, 116, 117,
 287

© Springer International Publishing Switzerland 2016
C. Burgess, B. Vis, *Interkosmos*, Springer Praxis Books, DOI 10.1007/978-3-319-24163-0

C

Carlos Ulloa Military school, 130
Carr, Gerald, 21
Carter, President Jimmy, 119
Casper, John, 209
Castro, Fidel, 127, 134, 136, 137, 139
Cazaux Flight Test Center, 270
Ceausescu, Elena, 175
Ceausescu, Nicolae, 176
Center National d'Études Spatiales (CNES), 186,
     190, 200–202, 267–270, 273, 281
Cernan, Gene, 29
Ceske Budejovice, Czechoslovakia, 13, 14
Chechina, Yelena, 270
Chrétien, Amy, 281
Chrétien, Emmanuel, 186
Chrétien, François, 186
Chrétien, Gregoire, 184
Chrétien, Jacques, 184
Chrétien, Jean-Baptiste, 186
Chrétien, Jean-Loup, 184, 185, 187, 188, 191,
     193, 199, 200, 203, 229
Chrétien, Jérôme, 184
Chrétien, Marie-Blanche, 184
Chrétien, Mary-Cathryn, 186, 281
Chrétien, Olivier, 186, 281
Chrétien, Philippe, 184
Chukhlov, Nikolay, 165
Cihlar, Jozef, 29
Clervoy, Jean-François, 268, 269
Coc, Nguyen Van, 115
College de l'Hay-les-Roses, 270
College Saint-Charles à Saint-Brieuc, 184
Comaneci, Nadia, 176
Cosmic Returns (book), 27
*Cosmosul-Laborator şi uzină pentru viitorul
     omenirii*, 180
Couette, Antoine, 268
Curien, Hubert, 190

D

Daily Mirror (Sydney newspaper), 260
Daily Telegraph (U.K. newspaper), 205
Dakov, Mako, 238
Danczak, Boleslaw, 38
Dandass, Flt. Lt. T. S., 209
Dauran, Alia Nur Mohammad, 262
Dauran, Mohammad, 253–258, 262
Davidova, Hana, 27
Davidova, Vaclav, 27
Deblin Air School, Poland, 37, 43, 53
Dediu, Dumitru, 163, 166, 168, 182
de Gaulle, President Charles, 184

d'Estaing, President Giscard, 186, 194
Deutsche Film-Aktiengesellschaft (DEFA), 73
Deutsches Museum, Munich, 76
Die Linke (online bulletin), 75
Dimensiuni psihice ale zborului aerospaţial, 180
Dimitrie Cantemir National College, 182
Dinh, Vu Ngoc, 111
Dobroslavtsi Air Base, Sofia, 79
Domin, Boris, 29
Dong, Pham Van, 124
Dorozynski, Janusz, 38
Duan, Le, 124
*Dumitri-Dorin Prunariu
     Biografia Unui Cosmonaut* (book), 182
Dunayev, Aleksandr, 238, 243, 252, 269
Dung, Van Tien, 124
Dvorak, Dr. Antonin, 15, 16, 61
Dzhanibekov, Vladimir, 21, 84, 99, 101, 102,
     147–154, 191, 193, 194, 199, 266,
     286–288
Dzhezkazgan, Kazakhstan, 71, 72, 105, 137, 153,
     174, 246, 280
Dzhurov, Chavdar, 80, 81
Dzhurov, Gen. Dobri, 80, 238

E

École de l'Air, 186, 188, 202, 270
École Sainte-Marie-Lebrun à Bordeaux, 188
Edwards Air Force Base, California, 201, 209
Elek, Laszlo, 95–97
Empire Test Pilots' School (ETPS), 189
Erhambaar (Mongolian candidate), 145
Euromir (space program), 74
European Astronaut Centre, 283
European Space Agency (ESA), 2, 74, 187, 190,
     250, 268, 289

F

Falcón, Daira, 130
Falcón, José Armando López, 129, 132
Falcón, Kintero Iraida, 130
Faris, Ghadil, 227
Faris, Gind Akil, 225, 234
Faris, Kutayb, 227
Faris, Mir, 234
Faris, Muhammed Ahmed, 224–229, 235, 270, 290
Farkas, Adam, 107
Farkas, Aida, 107
Farkas, Aniko, 97
Farkas, Bertalan, 92–98, 101–108, 138,
     168, 288
Farkas, Bertalan (son), 107

Farkas, Erzhebet, 93
Farkas, Lajos, 93
Fiolek, Kazimierz, 38
Flade, Klaus-Dietrich, 74
Fliegerkosmonauten (movie), 285
Furniss, Tim, 153

G

Gagarin Air Force Academy, 14, 16, 43, 112,
     114, 252
Galkin, Aleksandr, 227
Gandhi, Mahatma, 215
Gandhi, Prime Minister Indira, 206, 215, 217
Gankuyag, Maydarzhavyn, 146
Ganzorig, Maydarzhavyn, 145–147, 157
GDR Academy of Sciences, 68
Georgi Benkovski Higher Air Force School, 79–81
German Aerospace Center, 74
Gibson, Edward, 21
Gilmour, David, 276
Glavkosmos, 238, 243, 252, 269
Glazkov, Yuriy, 117
Glushko, Valentin, 170
Golbs, Eberhard, 60–62, 74, 75
Gombik, Stefan, 15
Goncharov, Igor, 232
Gorbachev, Mikhail, 233, 252, 266, 267, 276
Gorbatko, Viktor, 84, 101, 117–125, 288
Grabe, Ron, 209
Grass Near My Home, The (song), 92
Grechko, Georgiy, 18, 21, 25, 27, 83, 131,
     212, 286
Gretsov, Prof. Peter, 249
Gubarev, Aleksey, 18, 25, 27, 32, 116, 286
Guran, Cristian, 162
Gurragcha, Batbayar, 143
Gurragcha, Odbayar, 143
Gurragchaa, Zhugderdemidiyn, 143–147, 153,
     157, 158
Gutyina, Peter, 96

H

Haase, Dr. Hans, 61
Habib, Madyan, 227, 235
Habib, Munir Habib, 224–229, 234, 236
Habib, Rayed, 227, 235
Habib, Yumna, 235
Haigneré, Jean-Pierre, 268, 269
Hajduk, Jozef, 38
Halka, Henryk, 38–40, 61
Hall, Rex, 63, 85, 261

Harvey, Brian, 151
Hasan, Jan, 15
Haufe, Christian, 60
Hawthorne, Douglas, 52
Hazan, Barukh, 120
Hermaszewski, Aline, 36
Hermaszewski, Anna, 36
Hermaszewski, Boguslaw, 36, 37
Hermaszewski, Emilia, 37, 53
Hermaszewski, Emilia (daughter), 38
Hermaszewski, Kamila, 36
Hermaszewski, Miroslaw, 38
Hermaszewski, Miroslaw (Mirko), 36
Hermaszewski, Miroslaw (son), 36
Hermaszewski, Roman, 36
Hermaszewski, Sabina, 36
Hermaszewski, Teresa, 36, 37
Hermaszewski, Wladyslaw, 36, 37
Hermes program, 201
Hewad (Afghan newspaper), 255
Hideg, Dr. János, 96
Himmelsstürmer (film), 73
Hindon Air Force Station, Delhi, 209
Hooper, Gordon, 111, 242
Houtman, Maarten, 83, 286
Humboldt University, Berlin, 68
Hungarian Astronautical Society, 107
Hungarian National Defense Association, 93
Husak, Gustav, 25

I

Ichinkhorol, Chultemiin, 143
Indian satellites
     Aryabhata, 205
     Bhaskara, 205
     Bhaskara-2, 205
     IRS-1A, 205
Indian Space Research Organization (ISRO), 205
Interkosmos spacecraft (unmanned)
     Bolgaria-1300, 7
     Interkosmos-1, 5, 6
     Interkosmos 2, 5, 11
     Interkosmos-15, 7
     Interkosmos-18, 7
     Koronass-1, 9
     Kosmos-261, 4, 5
     Magion, 7, 8
     Oreol-3, 9
     Vertikal-1, 8
International Academy of Aeronautics, 283
International Space Station (ISS), 103, 108, 249,
     281, 283

Intersputnik program, 3, 12
Isaulov, Yuriy, 16, 17
Istoria aviaţiei române (book), 180
Ivanchenkov, Aleksandr, 47–49, 191, 193, 197,
    199, 266, 286, 287
Ivanov, Georgi, 77–81, 83–87, 89, 90, 92, 101,
    138, 238, 240, 244, 246, 285, 288
Ivanov, Natalia (née Rousanova), 79, 89

J
Jahid, Mohammad, 254
Jähn, Erika, 57, 58, 74
Jähn, Grit, 57
Jähn, Marina, 57, 65
Jähn, Paul, 57
Jähn, Sigmund, 58, 60–62, 64, 65, 72, 73, 75, 76,
    138, 285, 287, 288, 298
Jamal, Sahib, 252
Jan, Akar, 254
Jankowski, Aniele, 43
Jankowski, Katerina, 43
Jankowski, Zenon, 38–40, 42–44, 52, 61, 295, 296
Jehnichen, Walther, 60, 61
Jensen, Amy *See* Chrétien, Amy
Joban, Jean-Pierre, 187
Johnson Space Center, 260, 281
Johnson Space Center, Texas, 185, 283
José Antonio Echeverria Polytechnic University,
    130
Juin, Gerard, 187

K
Kafel, Mieczyslaw, 38
Kakalov, Anastasia, 77
Kakalov, Ani, 79
Kakalov, Georgi *See* Ivanov, Georgi
Kakalov, Ivan, 77, 80
Kaklov, Natalia, 79
Kaleri, Aleksandr, 270, 283
Kamanin, Nikolay, 77
Karl Marx Higher Institute of Economics, 240
Karol Swierczewski General Staff Academy, 37, 43
Karzai, President Hamid, 264
Kennedy Space Center, Florida, 30
Khan, Amer, 254
Khrunov, Yevgeniy, 131, 132, 141, 164
Khrushchev, Nikita, 235
Kidger, Neville, 269
Kiev Higher Air Force Engineering School, 252
Kilian Gyorgy Flight Technical College, 94, 95
Klet Observatory, Czechoslovakia, 31
Klima, Ladislav, 15, 16, 61

Klimuk, Pyotr, 44, 101, 116, 287
Klukowski, Krzusztof, 61
Köllner, Eberhard, 60–62, 74–76
Komarov, Vladimir, 116
Kopasz, Gyusi, 93
Kosice Air Force College, 16
Kotelnikov, Vladimir, 149, 267
Koval, Aleksandr, 213
Kovalyonok, Vladimir, 47, 70, 150, 151, 170, 191,
    287, 288
Kowal, Szymon, 38
Kozlekedesi Muzeum, Budapest, 108
Krasnodar Military Aviation Institute, USSR, 94
Krikalyov, Sergey, 270–272, 275–277
Kubasov, Valeriy, 44, 84, 98, 101, 102, 104,
    106, 107
Kuziora, Tadeusz, 38

L
La Cinci Minute după Cosmos, 180
Laveykin, Aleksandr, 229, 230, 232–234
Lazarov, Tsvetan, 78
20,000 Leagues Under the Sea (book), 78
Lebedev, Valentin, 120, 193, 194
L'École Communale à Ploujean, 184
Lenart, Jozef, 21
Leonov, Aleksey, 20, 26, 97, 101, 228, 229
Levlev-Sawyer, Clive, 243
Lewenko, Kazimierz, 38
Lewis, Frank, 112
Liem, Bui Thanh, 113, 115–117, 125, 168
Lions, Dr. Jacques-Louis, 267
Lobaina, Maria *See* Méndez, Maria
Loudon, Bruce, 205
Lyakhov, Vladimir, 85, 88, 102, 120, 147, 241,
    242, 247, 254–262, 280
Lycée Chaptal, 188
Lycée de Cachan, 270
Lycée de Morlaix, 184

M
Magyari, Bela, 93–97, 99, 101, 102, 107, 108
Magyari, Marthja, 107
Makarov, Oleg, 21, 286
Malhotra, Mira, 209, 215
Malhotra, Rakhi, 209
Malhotra, Ravish, 206, 208, 209, 211, 212, 215,
    216, 219, 220
Malhotra, Rohid, 209
Malinovskiy, Rodion, 77
Malyshev, Yuriy, 101, 190, 191, 212, 215, 217,
    219, 255

Marian, Boleski, 38
Masha (Russian mascot), 70
Mason, Nick, 276
Mayor, Federico, 203
McDow, Lt. Richard, 114
Medvedev, Dimitriy, 140
Melnikov, Ivan, 140
Men and Women of Space, 52
Méndez, Arnaldo Tamayo, 101, 127–129, 132,
        137, 138, 165, 168, 288, 290
Méndez, Esperanza, 127
Méndez, Rafael, 127
Méndez, Maria, 129
Menon, Narayanan, 209
Miller, Maj. R. C., 114
Minh, Ho Chi, 113, 119, 125
Misch, Peter, 60, 61
Mitterand, President François, 267, 271, 276
Moati, Nicole, 199
Mohammed, Khyal, 254
Mohmand, Abdul Ahad, 252, 253, 263, 264
Mohmand, Bibigul, 255
Mohmand, Hila, 255
Mongolian Academy of Sciences, 145, 157
Moscow News (newspaper), 18
Musée de l'Air, Paris, 306
Muska, Annie, 182
Muzeum Polskiej Techniki Wojskoweg, 54

N
Nahimov Navy School, Varna, 79
Najibullah (Najib), Gen. Secretary Mohammad,
        252
Nakov, Ivan, 80, 81
Nasser, Gamal Abdel, 235
National Defense Academy, India (NDA), 207, 208
National Military Heritage Museum, Bucharest, 182
Nehru, Pandit, 215
Nehru Planetarium, 222, 306
Neumann, Have, 96
Nikolayev, Andriyan, 147, 211
Nizam College, Hyderabad, 206
Novosti Kosmonavtiki (newspaper), 233

O
Oberg, James, 1, 27, 260
Oberth, Hermann, 180
October War Panorama Museum, 236
OKB-586 (design bureau), 7
Olteanu, Constantin, 175
Orbity Soyrudnichestva (book), 55
Otakara, King Premysla II, 13

P
Pandeliev, Velian, 88
Patrice Lamumba University, 163
Pavlik, Frantisek, 15
Pazur, Czeslaw, 39
Pelcak, Hannah, 16, 33
Pelcak, Martin, 33
Pelcak, Milos, 16
Pelcak, Oldrich, 15, 16, 18–20, 29, 32, 61
Pelcak, Oldrich (grandson), 33
Pelcak, Oldrich (son), 16
Petrov, Boris, 6, 145
Pink Floyd, 275, 276
Piso, Marius, 179
Plovdiv Museum of the Air, Bulgaria, 90
Pogue, William, 21
Polish Astronautical Society, 53
Polish Institute of Physics, 48
Polish Military Institute of Aviation Medicine, 41
Polyakov, Valeriy, 255, 256, 258, 259, 271, 275, 276
Popov, Leonid, 101, 102, 104, 119, 120, 122, 123,
        136, 165–168, 170, 172, 174, 175, 184, 288
Prague Institute of Aviation Medicine, 15
Prerov Air Base, 14
Prio, Carlos, 127
Prunariu, Crina, 159, 182
Prunariu, Dimitru-Dorin, 134, 159, 162, 182
Prunariu, Ovidiu-Daniel, 182
Prunariu, Radu-Catalin, 182
Prytanée Militaire de la Flèche, 188
Putin, Vladimir, 31

R
Radev, Kiril, 80, 81
Raj, Mulk, 217
Rakovski Military Academy, Sofia, 240
Rateb, Ahmed, 224
Raykov, Nikolay, 239
Reagan, President Ronald, 266
Rebeldi Technical Institute, 128
Red Star (newspaper), 1, 77
Red Star in Orbit (book), 1, 27
Red Star newspaper, 27
Reinhold, Wolfgang, 60
Remek, Anna, 27, 31
Remek, Blanka, 13, 16
Remek, Dana, 13
Remek, Jana, 31
Remek, Jitka, 13
Remek, Jozef, 13, 16
Remek, Vladimir, 11–13, 15, 16, 18, 21, 25,
        29–33, 101, 138, 285, 286, 290
Rêves d'étoiles (book), 274, 281

Rojek, Bogdan, 39
Roman, Petre, 178
Romanenko, Yuriy, 21, 25, 27, 83, 84, 88, 131,
    132, 136, 137, 141, 164, 166, 168, 229,
    230, 240
Romanian Space Agency, 178–180
Rossowski, Fr. Mieczyslaw, 36
Rousanova, Natalia *See* Kaklov, Natalia
Rozman, Gyula, 94
Rukavishnikov, Nikolay, 17–19, 83, 85, 88, 90,
    212, 240
Russia's cosmonauts, 92
Russia's Cosmonauts (book), 63, 96
Russian Space Probes (book), 151
Ryumin, Valeriy, 85, 86, 102, 104, 119–123, 136,
    147, 228, 288

S

Sajncog, Sanzaadambyn, 146, 147
Sandmännchen (puppet), 69
Saravsambuu (Mongolian candidate), 145
Sarvar, Mohammad, 252
Savinykh, Viktor, 170, 173, 226, 242, 244, 246,
    247, 288
Sedivy, Jiri, 29
Seleverstov, Maj. A., 280
Selong, Jaroslav, 30
Serebrov, Aleksandr, 90, 242, 247, 270, 271
Sharma, Devendraneth, 206
Sharma, Kapil, 211, 215, 221
Sharma, Kritika, 221
Sharma, Madhu, 207, 210, 211, 215, 221
Sharma, Mansi, 211
Sharma, Rakesh, 101, 206, 207, 211, 212, 214,
    215, 219–221, 290
Sharma, Tripta, 206
Shatalov, Vladimir, 19, 27, 116, 192, 219
Shayler, David J., 63, 153, 261
Shirendyb, Prof. B., 149
Shkola Orlyat High Aviation School, 53
Sigmund Jähn (ship), 73
Simonyi, Charles, 108
Singh, President Zail, 215
Smetanin, Nikolay, 280
Sofia Echo (newspaper), 243
Solovyev, Anatoli, 242, 244, 283
Solovyev, Anatoliy, 247
Soto, Dr. Luis Diaz, 129
Soviet Weekly (newspaper), 149, 173
Soyuz
    A Universal Spacecraft (book), 85, 261
Soyuz spacecraft
    Soyuz -1, 116

Soyuz -2, 116
Soyuz -4, 293
Soyuz -5, 131, 164, 293
Soyuz -6, 98, 216
Soyuz -7, 1, 117
Soyuz -10, 83
Soyuz -16, 83
Soyuz-17, 18
Soyuz -22, 116
Soyuz -24, 117
Soyuz -25, 86
Soyuz -26, 21, 83, 99, 131, 287
Soyuz -27, 21, 99, 147, 153, 286, 287
Soyuz -28, 20, 21, 23–28, 32–34, 84, 286, 287,
    295, 303, 305
Soyuz -29, 65, 70, 71, 76, 191, 287, 306
Soyuz -30, 45, 47, 286, 304, 306
Soyuz -31, 64–67, 70, 75, 76, 116, 287
Soyuz -32, 85, 102, 147, 287
Soyuz -33, 84–90, 92, 101, 102, 238, 240, 285,
    287, 290, 295
Soyuz -34, 101, 102, 287
Soyuz -35, 102, 105
Soyuz -36, 92, 99, 102–104, 106, 108, 120,
    122, 123, 288, 294
Soyuz -37, 114, 117, 119, 120, 122–125, 288
Soyuz -39, 304
Soyuz -40, 166, 168–171, 174, 175, 177, 179, 182
Soyuz T-2, 289
Soyuz T-3, 191, 289
Soyuz T-4, 150, 153, 170, 174, 225, 288, 289
Soyuz T-5, 289
Soyuz T-6, 191, 193, 195, 196, 198, 199, 289, 306
Soyuz T-7, 289
Soyuz T-9, 225
Soyuz T-10, 216, 219, 222, 289, 306
Soyuz T-11, 213, 215, 216, 289, 296, 297
Soyuz T-13, 225
Soyuz TM-2, 141, 230, 232–234
Soyuz TM-3, 225, 226, 230–233, 236, 270
Soyuz TM-4, 245, 246
Soyuz TM-5, 90, 242–244, 248, 259
Soyuz TM-6, 255, 256, 258, 271, 275, 278, 279
Soyuz TM-7, 203, 266, 268, 270–272, 274–276
Soyuz TM-8, 269
Soyuz TM-14, 74, 283
Soyuz TM-15, 283
Soyuz TMA-10, 103, 108
Soyuz TMA-14, 103, 108
Spiru Haret University, Budapest, 182
Stamenkov, Eugene, 77
Stamenkov, Karamfil, 77
Stamenkov, Krum, 77
Stamenkov, Stamenka, 77

Stampf, Olaf, 57
Star City, Moscow, 10, 15, 162–164, 166, 175
Stoyanov, Dobromir, 240
Stoyanov, Krasimir, 239–242, 247–249
Stoyanov, Lyudmila, 240
Stoyanov, Mikhaela, 240
Strekalov, Gennadiy, 213, 215
Stuban, Aniko *See* Farkas, Aniko
Syra-Juden, 254
Syrenhorloo, Darzaagijn, 145, 146
Szabó, Jozef, 95, 96
Szijjártó, Péter, 107

T
Tan, Tran Phuong, 112
Tasalov, Sergey, 242
TASS (Soviet newsagency), 23, 84, 85, 119, 193, 260
Tereshkova, Valentina (Valya), 79, 116
Tesentr Podgotovki Kosmonavtov (TsPK)
      See Yuri Gagarin Cosmonaut Training Center
Tho, Nguyen Huu, 124
Titov, German, 244
Titov, Vladimir, 244, 274, 276, 279
To a Friendly World (book), 154
Todor Kirkov Technical School, Bulgaria, 79
Tognini, Anne-Marie, 270
Tognini, Benedicte, 270
Tognini, Ginette, 270
Tognini, Jean, 270
Tognini, Michel, 229, 268, 271, 283
Tognini, Nicolas, 270
Tuan, Pham, 101, 110–119, 121, 123–125, 138,
    168, 288
Tuan, Pham Anh (son), 125
Tuan, Tkhu, 125
Tuyet, Nguyen Thi, 114
Twerdlowski, Jan, 47

U
Ulaan-Bataar Agriculture Institute, 143
Un Cubano en el Cosmos (book), 140
Un Roman Zboara Spre Stele (movie), 177
USSR Academy of Sciences, 3, 48, 149, 199, 267
U.S. space shuttle missions
    STS 51-D, 200
    STS 51-E, 200
    STS 51-G, 200
    STS-86, 281
    STS-93, 283
USSR Academy of Sciences, 3, 4, 48, 149,
    199, 267

USSR cosmodromes
    Baykonur (Kazakhstan), 84
    Kapustin Yar (Russia), 5, 205
    Plesetsk (Russia), 4, 9
U.S. space shuttles
    Atlantis, 281
    Challenger, 243
    Columbia, 150, 283
    Discovery, 200, 266
USSR space stations
    Mir, 75, 90, 203, 227, 229, 232, 233, 238
    Salyut-4, 90
    Salyut-5, 117
    Salyut-6, 10, 21, 23, 25, 26, 43, 47–50, 54, 65,
        67, 68, 70, 84, 85, 88, 89, 103–105, 114,
        116, 119, 120, 123, 136, 137, 142, 184, 191
    Salyut-7, 90, 184, 190, 192–197, 216, 218,
        224, 225, 266, 268
    Salyut-10, 90
    Salyut-29, 47
USSR unmanned spacecraft, 141
    Buran, 273, 274, 281
    Granat, 250
    Phobos-1, 250
    Phobos-2, 250
    Progress-9, 104
    Progress-10, 120
    Progress-36, 242

V
Varnier, Françoise, 187
Venkatarman, Ramaswamy, 215
Verne, Jules, 78
Vietnam News (newspaper), 110, 125
Vietnam People's Museum, Hanoi, 113
Viktorenko, Aleksandr, 225, 226, 228, 232–234,
    270
Volkov, Aleksandr, 268, 269, 271, 272, 276
Volynov, Boris, 101, 211
Vondrousek, Michael, 61
Vorobyev, Lt. Col. N., 280
Vorontsov, Yuliy, 205
Voroshilov, K.E. Military Academy, 263
Vostok manned spacecraft, 116
    Vostok-6, 116
    Vostok-5, 116
Vyssi Letecke Uciliste (flight college), 14

W
Waliszkiewicz, Jan, 39
Warumzer, Jakub, 36

Watanjar, Mohammad, 252
Wehner, Wolfgang, 60
Weigel, Endre, 96
White Sun of the Desert (movie), 92
Wroclaw Aeroclub, 37
Wrzesien, Henryk, 39

Y
Yeisk Higher Military Aviation School, 129
Yeliseyev, Aleksey, 86
Yen Bai Airport, 112
Yen, Nguyen Thi, 113, 114
Yotov, Ventsislav, 80
Youchev, Georgi, 81

Yribar, Torres, 129
Yuriy Gagarin Cosmonaut Training Center, 162,
    186, 210, 224, 242

Z
Zahariev, Zahari, 77
Zakutnyana, Olga, 151
Zamin, Shere, 254
Zaytsev, Andrey, 242
Zemlyane (band), 92
Zhivkov, Todor, 243, 246
Zhukovskiy Air Force Engineering
    Academy, 81, 143
Zvyozdnyy Gorodok, 189